OLDSMOBILE TORONADO 1966-1978

ISBN 9780946489879

Brooklands Books
BROOKLANDS BOOKS LTD.
P.O. BOX 904, AMERSHAM,
BUCKS. HP6 9JA. UK

OLDT www.brooklandsbooks.com 5W5/2735

BROOKLANDS BOOKS - ROAD TEST

www.brooklandsbooks.com

Abarth Road Test Portfolio 1950-1971
AC Cars 1904-2011
AC Ace Aceca Greyhound Road Test Portfolio 1953-62
AC & Cobra 1962-2011
Alfa Romeo A Brooklands Portfolio 1920-1933
Alfa Romeo A Brooklands Portfolio 1934-1940
Alfa Romeo A Brooklands Portfolio 1920-1940
Alfa Romeo Giulietta Gold Portfolio 1954-1965
Alfa Romeo Giulia Berlina Lim. Edit. Extra 1962-1976
Alfa Romeo Giulia Coupes Lim. Edit. Ultra 1973-1976
Alfa Romeo Alfasud 1972-1984
Alfa Romeo Alfetta Gold Portfolio 1972-1987
Alfa Romeo Alfetta GTV6 1980-1986
Alfa Romeo Spider Ultimate Portfolio 1966-1994
Alfa Romeo Spider & GTV Perf. Port. 1995-2005
Allard Limited Edition Ultra
Alpine Renault Ultimate Portfolio 1958-1995
Alvis Gold Portfolio 1919-1967
AMC Rambler Limited Edition Extra 1956-1969
AMX & Javelin Gold Portfolio 1968-1974
Aston Martin Gold Portfolio 1921-1947
Aston Martin Ultimate Portfolio 1948-1968
Aston Martin Ultimate Portfolio 1968-1980
Aston Martin Ultimate Portfolio 1981-1993
Aston Martin Ultimate Portfolio 1994-2006
Auburn Cord Duesenberg - A Brooklands Portfolio
Audi Quattro Gold Portfolio 1980-1991
Audi TT Performance Portfolio 1998-2006
Austin Seven & Ten Road Test Portfolio
Austin-Healey 100 & 100/6 Gold Port. 1952-1959
Austin-Healey 3000 Road Test Portfolio
Austin-Healey Sprite Bug-Eye Mk 1 Road Test Portf.
Avanti Limited Edition Extra 1962-1991
Bentley & Rolls-Royce Portfolio 1990-2002
Berkeley Sportscars Limited Edition
BMW 700 Limited Edition 1959-1965
BMW 2002 Ultimate Portfolio 1968-1976
BMW 6 Cylinder Coupes & Saloons Gold P. 1969-76
BMW 320, 323, 325 (6 cyl.) Gold Port. 1977-1990
BMW 3 Series Gold Portfolio 1991-1997
BMW M3 Ultimate Portfolio 1986-2006
BMW M5 Gold Portfolio 1980-2003
BMW 5 Series Gold Portfolio 1988-1995
BMW 6 Series Ultimate Portfolio 1976-1989
BMW 7 Series Performance Portfolio 1986-1993
BMW 8 Series Performance Portfolio
BMW X5 Limited Edition Extra 1999-2006
BMW Alpina Performance Portfolio 1967-1987
BMW Alpina Performance Portfolio 1988-1998
BMW Z3, M Coupe & M Roadster Gold Port. 1996-02
Bond Cars 1949-1974 Road Test Portfolio
Borgward Isabella Limited Edition
Bricklin Golf Portfolio 1974-1975
Bristol Cars Portfolio
Bugatti Type 10 to Type 40 Road Test Portfolio
Bugatti Type 41 to Type 55 Road Test Portfolio
Bugatti Type 57 to Type 251 Road Test Portfolio
Bugatti Type 10 to Type 251 Road Test Portfolio
Buick Performance Portfolio 1947-1962
Buick Muscle Portfolio 1963-1973
Buick Riviera Performance Portfolio 1963-1978
Cadillac Performance Portfolio 1948-1958
Cadillac Performance Portfolio 1959-1966
Cadillac Eldorado Performance Portfolio 1967-1978
Cadillac Allante Limited Edition Extra
Caterham Seven Road Test Portfolio 1974-1999
Caterham Seven Road Test Portfolio 2000-2010
Checker Automobiles Road Test Portfolio
Chevrolet Impala & SS Muscle Portfolio 1958-1972
Corvair Performance Portfolio 1959-1969
Chevy II & Nova SS Gold Portfolio 1962-1974
Chevrolet Chevelle & SS Gold Portfolio 1964-1972
Camaro Muscle Portfolio 1967-1973
Blazer & Jimmy Limited Edition Extra 1969-1982
Blazer & Jimmy Limited Edition Extra 1983-1994
Camaro Performance Portfolio 1993-2000
Chevrolet Caprice Road Test Portfolio 1965-1990
Chevrolet Corvair Performance Portfolio 1959-1965
Chevrolet Corvette Gold Portfolio 1953-1962
Chevrolet Corvette Sting Ray Gold Port. 1963-1967
Chevrolet Corvette Gold Portfolio 1968-1977
Chrysler Imperial Gold Portfolio 1951-1975
Chrysler 300 Gold Portfolio 1955-1970
Citroen Traction Avant Limited Edition Premier
Citroen 2CV Ultimate Portfolio 1948-1990
Citroen DS & ID 1955-1975
Citroen DS & ID Gold Portfolio 1955-1975
Citroen CX Road Test Portfolio
Citroen SM Limited Edition 1970-1975
Crosley & Crosley Specials Limited Edition
Cunningham Automobiles 1951-1955
Daimler SP250 Dart & V-8 2.5 litre 250 RTP 1959-69
Datsun Roadsters Performance Portfolio 1960-71
Datsun 240Z & 260Z Road Test Portfolio
Datsun 280Z & 280ZX Road Test Portfolio
Delahaye Road Test Portfolio
Delage Road Test Portfolio
DeLorean Gold Portfolio 1977-1995
De Soto Limited Edition 1952-1960
De Tomaso Collection No. 1 1962-1981
Dodge Limited Edition 1949-1959
Dodge Dart Limited Edition Extra 1960-1976
Dodge Muscle Portfolio 1964-1971
Charger Muscle Portfolio 1966-1974
Edsel Limited Edition 1957-1960
Excalibur Collection No. 1 1952-1981

Elva Sports Racers Road Test Portfolio
ERA Gold Portfolio 1934-1994
Facel Vega Limited Edition Extra 1954-1964
Ferrari Road Cars 1946-1956 Road Test Portfolio
Ferrari Dino Limited Edition Extra 1965-1974
Ferrari 308 & Mondial Ultimate Portfolio 1975-85
Ferrari 328 348 Mondial Ultimate Portfolio 1986-94
Ferrari F355 & 360 Gold Portfolio 1995-2004
Fiat 500 1936-1972 Road Test Portfolio
Fiat 600 & 850 Gold Portfolio 1955-1972
Fiat Dino Road Test Portfolio
Fiat 124 Spider Performance Portfolio 1966-1985
Fiat Abarth 1972-1987 Road Test Portfolio
Fiat X1/9 Gold Portfolio 1973-1989
Fiat Barchetta Road Test Portfolio
Ford Zephyr, Zodiac, Executive Mk. III & IV 1962-71
High Performance Capris Gold Portfolio 1969-1987
Capri Muscle Portfolio 1974-1987
High Performance Fiestas 1979-1991
Ford Escort RS & Mexico Limited Edition 1970-1979
Ford Thunderbird Performance Portfolio 1955-1957
Ford Thunderbird Performance Portfolio 1958-1963
Ford Thunderbird Performance Portfolio 1964-1976
Ford Fairlane 1955-1970
Ford Fairlane Performance Portfolio 1955-1970
Ford Edsel 1957-1960 Road Test Limited Edition
Ford Ranchero Muscle Portfolio 1957-1979
Ford Falcon Performance Portfolio 1960-1970
Ford Galaxie and LTD Gold Portfolio 1960-1976
Ford GT40 & GT Ultimate Portfolio 1964-2006
Ford Bronco 4x4 Performance Portfolio 1966-1977
Ford Bronco 1978-1988
Ford Torino Performance Portfolio 1968-1974
Shelby Mustang Ultimate Portfolio 1965-1970
Mustang Muscle Portfolio 1967-1973
High Performance Mustang IIs 1974-1978
Mustang 5.0L Muscle Portfolio 1982-1993
Ginetta Cars Limited Edition Ultra 1958-2007
Hispano-Suiza Road Test Portfolio
Holden 1948-1962
Honda S500, S600, S800 Limited Edition
Honda - Acura NSX Ultimate Portfolio 1989-2005
Honda S2000 Performance Portfolio 1999-2008
Hudson Performance Portfolio 1946-1957
International Scout Gold Portfolio 1961-1980
Isetta Gold Portfolio 1953-1964
ISO & Bizzarrini Limited Edition Ultra 1962-1974
Jaguar XK120 XK140 XK150 Gold Portfolio
Jaguar Mk 7, 8, 10 & 420 Road Test Portfolio
Jaguar Mk 1 & Mk 2 1955-1969 Road Test Portfolio
Jaguar E-Type Ultimate Portfolio 1961-1975
Jaguar XJ6 Series I & II Gold Portfolio 1968-1979
Jaguar XJ6 Series III Perf. Portfolio 1979-1986
Jaguar XJ-S V12 Ultimate Portfolio 1988-1996
Jaguar XK8 & XKR Performance Portfolio 1996-2005
Jeep CJ-5 Limited Edition 1960-1975
Jeep CJ-5 & CJ-7 4x4 Perf. Portfolio 1976-1986
Jeep Wagoneer Performance Portfolio 1963-1991
Jeep J-Series Pickups 1970-1982
Jeepster & Commando Limited Edition 1967-1973
Jeep Cherokee & Comanche Pickups P. P. 1984-1991
Jeep Wrangler 4x4 Performance Portfolio 1987-99
Jeep Cherokee & Grand Cherokee 4x4 P. P. 1992-98
Jensen Road Test Portfolio 1934-1965
Jensen Interceptor Ultimate Portfolio 1966-1992
Jensen - Healey Road Test Portfolio 1972-1976
Jowett Road Test Portfolio
Kaiser - Frazer Limited Edition 1946-1955
Lagonda Gold Portfolio 1919-1964
Lancia Aurelia & Flaminia Gold Portfolio 1950-1970
Lancia Fulvia Gold Portfolio 1963-1976
Lancia Montecarlo & Scorpion 1975-1982 RTP
Lancia Stratos Limited Edition Extra
Lancia Delta & integrale Ultimate Portfolio
Land Rover Series I, II & IIA Gold Portfolio 1948-1971
Land Rover 90 110 Defender Gold Portfolio 1983-94
Lamborghini Performance Portfolio 1964-1976
Lamborghini Performance Portfolio 1977-1989
Lamborghini Gold Portfolio 1990-2004
Lincoln Gold Portfolio 1949-1960
Lincoln Continental Performance Portfolio 1961-69
Lincoln Continental 1969-1976
Lola Sports Racing Cars Limited Ed. Premier 1958-85
Lotus Sports Racers Portfolio 1951-1965
Lotus Seven Gold Portfolio 1957-1973
Lotus Elite Limited Edition 1957-1964
Lotus Elan Ultimate Portfolio 1962-1974
Lotus Cortina Road Test Portfolio
Lotus Elite & Eclat Road Test Portfolio 1984-1972
Lotus Excel Road Test Portfolio
Lotus Europa Gold Portfolio 1966-1975
Lotus Elise Limited Edition 1996-1999
Lotus Elise & Exige Gold Portfolio 1996-2005
Marcos Coupés & Spyders Gold Portfolio 1960-1997
Maserati Cars Performance Portfolio 1957-1970
Maserati Cars Performance Portfolio 1971-1982
Maserati Cars Performance Portfolio 1982-1998
Maserati Cars Ultimate Portfolio 1999-2007
Matra Road Test Portfolio
Mazda Miata MX-5 Performance Portfolio 1989-1997
Mazda Miata MX-5 Performance Portfolio 1998-2005
Mazda RX-7 Gold Portfolio 1979-1992
McLaren F1 • GTR • LM Sportscar Perf. Portfolio

Mercedes-Benz 1925-1939 - A Portfolio
Mercedes 190SL 300SL 300SLR - A Portfolio
Mercedes S & 600 Limited Edition Extra 1965-1972
Mercedes S Class Limited Edition Extra 1980-1991
Mercedes 230SL 250SL 280SL Ultimate Port. 63-71
Mercedes-Benz SLs & SLCs Ultimate Port. 1971-1989
Mercedes SLs Performance Portfolio 1989-1994
Mercedes E-Class W124 Road Test Portfolio 1986-95
Mercedes G-Wagen Gold Portfolio 1981-2005
Mercedes AMG Gold Portfolio 1983-1999
Mercedes AMG Ultimate Portfolio 2000-2006
Mercury Gold Portfolio 1947-1966
Mercury Comet & Cyclone Lim. Edit. Extra 1960-1975
Mercury Cougar Muscle Portfolio 1967-1973
Messerschmitt Gold Portfolio 1954-1964
MG Gold Portfolio 1929-1939
MG TD & TF Gold Portfolio 1949-1955
MG Y Type & Magnette Road Test Portfolio
MGA & Twin Cam Gold Portfolio 1955-1962
MG Midget Road Test Portfolio 1961-1979
MGB Roadsters 1962-1980
MGB MGC & V8 Gold Portfolio 1962-1980
MGF & TF Performance Portfolio 1995-2005
Mini Gold Portfolio 1959-1969
Mini Gold Portfolio 1969-1980
Mini Gold Portfolio 1981-1997
High Performance Minis Gold Portfolio 1960-1973
Mini Cooper Gold Portfolio 1961-1971
Mini Moke Ultimate Portfolio 1964-1994
Starion & Conquest Performance Portfolio 1982-90
Mitsubishi 3000GT & Dodge Stealth P.P. 1990-1999
Morgan Three-Wheelers Ultimate Portfolio 1909-52
Morgan Four-Wheelers Ultimate Portfolio 1936-67
Morgan Ultimate Portfolio 1968-1990
Morgan Ultimate Portfolio 1991-2009
Morris Minor MM & Series II Road Test Portfolio
Morris Minor 1000 Road Test Portfolio
Nash Limited Edition Extra 1949-1957
Nash-Austin Metropolitan Gold Portfolio 1954-1962
Nissan 350Z Road Test Portfolio
Nissan Skyline GT-R Ultimate Portfolio 1969-2010
Nissan Skyline GT-R Limited Edition Extra 1989-2002
Noble Sports Cars Road Test Portfolio
NSU Ro80 Limited Edition
Oldsmobile 1948-1963 Limited Edition Premier
Oldsmobile Cutlass & 4-4-2 Muscle Port. 1964-1974
Oldsmobile Muscle Cars 1964-1971
Oldsmobile Toronado 1966-1978
Opel GT Ultimate Portfolio 1968-1973
Opel Manta Limited Edition 1970-1975
Packard Automobiles 1920-1958 - A Portfolio
Pantera and Mangusta 1969-1974
Pantera Ultimate Portfolio 1970-1995
Panther Gold Portfolio 1972-1990
Plymouth Limited Edition 1950-1960
Plymouth Fury Limited Edition Extra 1956-1976
Barracuda Muscle Portfolio 1964-1974
Plymouth Muscle Portfolio 1964-1971
Pontiac 1946-1963 Limited Edition Premier
High Performance Firebirds 1982-1988
Firebird & Trans Am Performance Portfolio 1993-00
Pontiac Fiero Performance Portfolio 1984-1988
Porsche Sports Racing Cars UP 1952-1968
Porsche 917 • 935 • 956 • 962 Gold Portfolio
Porsche 365 Ultimate Portfolio 1952-1965
Porsche 911 1965-1969
Porsche 911 SC & Turbo Gold Portfolio 1978-1983
Porsche 911 Ultimate Portfolio 1990-1997
Porsche 911 Ultimate Portfolio 1998-2004
Porsche 912 Limited Edition Extra 1965-1976
Porsche 914 Ultimate Portfolio
Porsche 924 Gold Portfolio 1975-1988
Porsche 928 Gold Portfolio 1977-1995
Porsche 928 Takes On The Competition
Porsche 944 Ultimate Portfolio
Porsche 968 Limited Edition Extra
Porsche Boxster Ultimate Portfolio 1996-2004
Railton & Brough Superior Gold Portfolio 1933-1950
Range Rover Gold Portfolio 1970-1985
Range Rover Gold Portfolio 1985-1995
Range Rover Performance Portfolio 1995-2001
Riley RM Series Pathfinder 2.6 A Brooklands Port.
Rolls-Royce Silver Cloud & Bentley S Ultimate Port.
Rolls-Royce Silver Shadow Ultimate Port. 1965-80
Rover P4 1949-1959
Rover 3 & 3.5 P5 Gold Portfolio 1958-1973
Rover P6 1963-1977
Saab Sonett Collection No. 1 1966-1974
Studebaker Ultimate Portfolio 1946-1966
Studebaker Hawks & Larks Limit. Edit. Premier 56-66
Shelby Cobra Gold Portfolio 1962-1969
Subaru Impreza WRX Performance Port. 2001-2005
Sunbeam Alpine and Tiger Gold Portfolio 1959-1967
Sunbeam Alpine Limited Edition Extra 1959-1968
Sunbeam Tiger Limited Edition Extra 1964-1967
Suzuki SJ Gold Portfolio 1971-1997
Tatra Cars Road Test Portfolio
Toyota Land Cruiser Gold Portfolio 1956-1987
Toyota Land Cruiser 1988-1997
Toyota Supra Performance Portfolio 1982-1998
Toyota MR2 Road Test Portfolio 1984-1989
Toyota MR2 Road Test Portfolio 1990-1999
Toyota MR2 Road Test Portfolio 2000-2007

Triumph TR2 - TR3 1952-1960
Triumph TR2 & TR3 Gold Portfolio 1952-1961
Triumph TR4, TR5, TR250 1961-1968
Triumph TR6 Road Test Portfolio
Triumph TR7 & TR8 1975-1982
Triumph TR7 TR8 Gold Portfolio 1975-1982
Triumph Herald 1959-1971
Triumph Vitesse 1962-1971
Triumph 2000, 2.5, 2500 1963-1977
Triumph GT6 Gold Portfolio 1966-1974
Triumph Spitfire Road Test Portfolio
Triumph Stag Road Test Portfolio
TVR Limited Edition Ultra 1958-1985
TVR Performance Portfolio 1986-1994
TVR Performance Portfolio 2000-2005
VW Beetle Gold Portfolio 1935-1967
VW Beetle Gold Portfolio 1968-1991
VW Bus Camper Van Perf. Portfolio 1954-1967
VW Bus Camper Van Perf. Portfolio 1968-1979
VW Bus Camper Van Perf. Portfolio 1979-1991
VW Karmann Ghia Gold Portfolio 1955-1974
VW Golf Cabriolet 1979-2002 Road Test Portfolio
VW Corrado Limited Edition Premier 1989-1995
Volvo PV444 & PV544 Perf. Portfolio 1945-1965
Volvo 120 Amazon Ultimate Portfolio
Volvo 1800 Ultimate Portfolio 1960-1973
Volvo 140 & 160 Series Gold Portfolio 1966-1975
Westfield Performance Portfolio 1982-2004

MILITARY VEHICLES
Complete WW2 Military Jeep Manual
Dodge WW2 Military Portfolio 1940-1945
German Military Equipment WW2
Hail To The Jeep
Combat Land Rover Portfolio No. 1
Land Rover Military Portfolio
Off Road Jeeps Civilian & Military 1944-1971
US Military Vehicles 1941-1945
Standard Military Motor Vehicles-TM9-2800 (WW2)
VW Kubelwagen Military Portfolio 1940-1990
WW2 Allied Vehicles Military Portfolio 1939-1945
WW2 Jeep Military Portfolio 1941-1945

RACING
Can-Am Racing 1966-1969
Can-Am Racing 1970-1974
Can-Am Racing Cars 1966-1974
The Carrera Panamericana Mexico - 1950-1954
Le Mans - The Bentley & Alfa Years - 1923-1939
Le Mans - The Jaguar Years - 1949-1957
Le Mans - The Ferrari Years - 1958-1965
Le Mans - The Ford & Matra Years - 1966-1974
Le Mans - The Porsche Years - 1975-1982
Le Mans - The Porsche & Jaguar Years - 1983-91
Le Mans - The Porsche & Peugeot Years - 1992-99
Mille Miglia - The Alfa & Ferrari Years - 1927-1951
Mille Miglia - The Ferrari & Mercedes Years - 1952-57
Targa Florio - The Ferrari & Lancia Years - 1948-54
Targa Florio - The Porsche & Ferrari Years - 1955-64
Targa Florio - The Porsche Years - 1965-1973
Brabham Ralt Honda The Ron Tauranac

RESTORATION & GUIDE SERIES
BMW 2002 - A Comprehensive Guide
BMW 02 Restoration Guide
BMW E30 - 3 Series Restoration Bible
BMW E36 - 3 Series Restoration Tips & Techniques
BMW - 5 & 6 Series Restoration Tips & Techniques
Classic Camaro Restoration
Land Rover Restoration Portfolio
PC on Land Rover Series I Restoration
Lotus Elan Restoration Guide
Lotus Twin Cam Engines
MG T Series Restoration Guide
MGA Restoration Guide
PC on MGB Restoration
Mustang Restoration Tips & Techniques
Practical Gas Flow
Range Rover - The First Fifty
Restoring Sprites & Midgets an Enthusiast's Guide
Solex Carburettors Tuning Tips and Techniques
SU Carburetters Tuning Tips & Techniques
The Great Classic Muscle Cars Compared
Weber Carburettors Tuning Tips and Techniques

ENGINE SERIES
Chevy 265 & 283 — Chevy 302 & 327
Chevy 348 & 409 — Chevy 350 & 400
Chevy 396 & 427 — Chevy 454 thru 512
Chrysler Hemi — Chrysler 273, 318, 340 & 360
Chrysler 361, 383, 400, 413, 426 & 440
Ford 289, 302, Boss 302 & 351W — Ford Small Block
Ford 351C & Boss 351 — Ford Big Block

ROAD & TRACK SERIES
Road & Track on Auburn Cord & Duesenberg 52-84
Road & Track on Austin Healey 1953-1970
R & T on Cobra, Shelby & Ford GT40 1962-1992
Road & Track on Triumph Sports Cars 1953-1967
Road & Track on Triumph Sports Cars 1967-1974
Road & Track on Triumph Sports Cars 1974-1982
Road & Track - Henry Manney at Large and Abroad
Road & Track - Peter Egan Side Glances 1983-1992
Road & Track - Peter Egan Side Glances 1992-1997
Road & Track - Peter Egan Side Glances 1998-2002
Road & Track - Peter Egan Side Glances 2002-2006
Road & Track Best of PS

Contents

ACKNOWLEDGEMENTS

The Oldsmobile Toronado has already become a sought after "collectable" by discerning motorists in the US and there is no doubt that is will eventually become a "classic" due to its innovative mechanical features, styling and performance.

Front wheel drive has been with us for many years and both Alvis and Cord come immediately to mind. However, it was the French with the Traction Avant Citroens of the mid 30s who put it into serious production and made it commercially viable. More recently in 1959 the British launched their miniscule Minis and subsequently the Mini-Coopers which demonstrated the sporting potential of this layout by winning a cluster of rallies including the Monte Carlo 4 times. The Toronado takes up the mantle for the US where the Cord left off.

Our books are produced in small numbers for owners and potential owners of interesting vehicles. We hope that by making available once again the road tests and other articles relating to these cars, it will help to make their preservation easier and move rewarding. We are fortunate that the world's leading publishers sympathise with our aims and support our reference series by allowing us to include their informative copyright articles.

We are indebted in this instance to the management of *Autocar, Car Classics, Car and Driver, Car Life, Motor Sport, Motor Trend, Road Test, Road & Track, Sports Car Graphic* and *Wheels* for their generosity and understanding.

R.M. Clarke

The big news from Detroit for 1966

OLDSMOBILE TORONADO

BY JOHN R. BOND

THE TORONADO, Oldsmobile's new luxury car, is the big news for 1966, primarily because it was designed in response to the demand for a big roomy 6-passenger sedan that handles as near like a sports car as practicable.

The basic design problem with conventional cars having the engine forward driving the rear wheels is, and always has been, the ofttimes drastic change in fore and aft weight distribution. If, for example, a certain car has 55% of its weight forward, it may change to 55% at the rear with full load. If the designer elects to provide moderate understeer with one or two passengers the car may very well oversteer when fully loaded. In other words most conventional cars have too much under-

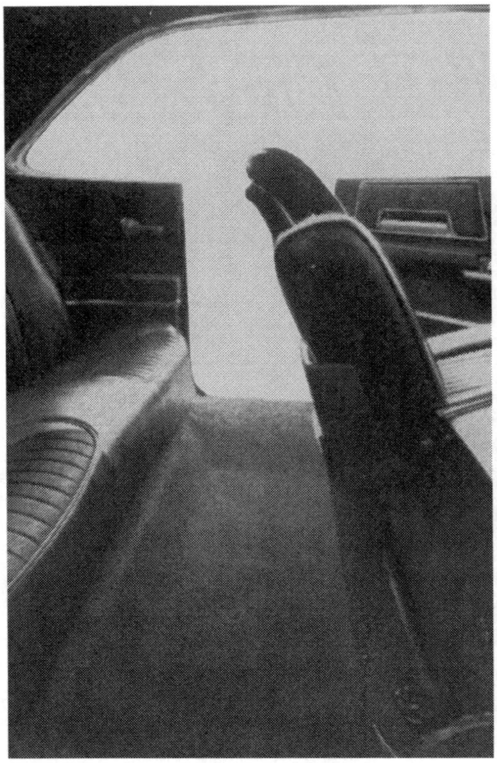

A major accommodation advantage with fwd is the lack of driveshaft tunnel to interfere with middle passengers' foot room.

OLDSMOBILE TORONADO

steer when lightly loaded, a compromise necessary to avoid oversteer when carrying a full load.

To get consistent handling qualities, Olds engineers went back to the drawing boards and came up with a front wheel drive vehicle that embodies some radical concepts. Among these are no less than 60.3% of the total weight on the front wheels, exceptionally stiff springing by American standards and quick steering—all propelled by a massive 7-liter engine developing 385 bhp. This is not to imply that a Toronado with the same power-to-weight ratio as a Corvette would beat same around the Elkhart Lake circuit—but it might be interesting to watch.

In short, like all cars designed for everyday use by Mr. & Mrs. Average there are some compromises in the TFD (Toronado Front Drive). Compromises, in our opinion, include the fact that the Toronado is rather large, surprisingly heavy and not exactly cheap to manufacture. But despite a massive preponderance of weight forward, the understeer is surprisingly moderate up to the limit of adhesion and the ride is still very good by boulevard standards. Traction in the wet, or in snow and ice, is much better than conventional rwd cars though we would suspect that a VW or a Corvair could climb a steeper snow-covered hill. Incidentally, there's ample clearance for chains on the front wheels.

As mentioned above, one serious compromise in the Toronado is its cost of manufacture. This is a specialty car, designed primarily to compete with the Buick Riviera and the Ford Thunderbird. We asked Olds engineers what the extra cost of fwd was but got only "no comment." However, most guesses on this subject put the extra cost at a minimum of $400 at retail, probably more. Nevertheless, the Toronado is undoubtedly GM's best answer to date on building a safer car for all types of roads and weather conditions. It's unfortunate that it costs so much more to produce.

Technically the TFD abounds in new and interesting features. Their first test car on the road appeared in 1959 and it was an F-85 with aluminum V-8 set transversely over the front wheels—with no right-angle drive gears. Traction was inadequate and the helical drive gears were noisy.

Hence the decision to use a large cast iron engine and chain drive for the transfer case. The Turbo-Hydra-Matic transmission is split into two separate parts; the turbo or converter portion bolts to the cylinder block in the normal position. The 3-speed automatic gearbox portion and the differential assembly are turned completely around and lie along the left hand side of the block. This centers about 950 lb of engine and drive train directly over the front driving wheels. The alternative Cord, Citroen, etc., arrangement of putting the engine slightly aft of the transaxle doesn't put as much weight forward and the engine proper tends to intrude on passenger space. Of course, if front end overhangs continue to increase we might possibly see a V-8 hanging out in front like the Saab and Ford Taunus 12-M. But as it stands now the Olds Toronado layout is a good solution despite the fact that the arrangement looks as if it's doing it the hard way.

The Toronado is about the size of an Impala with a wheelbase of 119 in. and an overall length of 211 in. It is surprisingly heavy at 4500 lb, about the same as the 126-in. Olds 98. The 2-door body seats six adults comfortably and the passenger in the middle will appreciate full springing underneath and full leg room in front. The doors are the widest in the industry to facilitate rear seat entry and exit. There are no windwings and this plus a unitized engine/drive-line package completely self-contained in the forward compartment makes the TFD one of the quietest cars on the road today. Perhaps it's quieter than a Rolls-Royce—or even a Ford.

The engine is a standard 425-cu-in. model with a special high-lift camshaft, larger diameter lifters, bigger intake ports in the head and 2.06-in. intake valves. It is rated by Olds at 385 bhp at 4800 rpm which, at only 10 bhp more than the previous top engine, seems somewhat conservative. Certainly this engine delivers at least 300 bhp to the driving wheels, perhaps as much as 325. It weighs 642 lb, by the way, and 879 lb with converter, drive chain and transmission, but no differential.

The engine is located 1.82 in. to the right and virtually dead center longitudinally over the front wheel centerline. Behind the engine is a normal GM variable-pitch torque converter with stall torque ratios of 1.8 to 2.2:1. This is housed in an enormous die casting which extends across and down to include an integral chain case and 3-speed Hydra-Matic transmission.

The chain itself is the old-fashioned silent type that used to be so popular for timing chains. It is 46.5 in. long and because it transmits up to 1045 lb-ft of torque (engine x 2.2) the width is 2 in. A great deal of engineering time and effort went into this chain drive and it is said to be absolutely silent and troublefree for the life of the car. There are no anti-whip devices, no adjustments. The driving sprocket has a rubber cushioned hub to damp out torsional vibration and lubrication is by splash.

The crankshaft centerline lies 17.29 in. above and parallel to the ground line. The chain case itself is on 11.062-in. centers and drops the drive 4.43 in. so that it coincides with the wheel center height thus eliminating the need for hypoid gears in the differential. Each sprocket is 7.75-in. diameter, has 65 teeth for low tooth loading and, of course, the drive ratio is 1:1.

The internals of the 3-speed unit are all standard GM parts except at the output or forward end. There's no tail-shaft extension; instead an iron differential housing is bolted on. This assembly is a snug fit alongside the crankcase and contains a very large ring gear, a straddle-mounted pinion and a compact planetary-type differential. The gears are spiral bevels, not hypoid, as mentioned earlier. Both the ring and pinion are of slightly smaller pitch than normal with 45 teeth on the ring gear, 14 on the pinion (where 9 to 11 teeth are more common). But the most unusual feature is planetary gears for the differential action. This was necessary, so we were told, because development testing indicated that limited slip devices were actually quite harmful to the handling qualities and even conventional bevel gear type differentials caused some feedback because of their high internal friction (when cornering). An additional advantage of the planetary gears comes from their lateral compactness which, in turn, allows a compact differential housing and a slightly longer left hand half-shaft than would otherwise be possible. All differential bearings are pre-loaded Timken tapered-rollers of very large size. Only one gear ratio is supplied: 3.214:1.

Drive for the right-hand front wheel passes under the shallow engine oil pan to a ball-type steadying bearing bolted on the right side of the block. The two half-shafts are of equal length and necessarily rather short. The right-hand shaft is actually three pieces. The first is the portion under the oil pan. While the distance between universal joint centers is the same as on the left hand side the right shaft proper is in two sections with a large rubber cushion joining

Under-hood view looks pretty much like average current U.S. car, but view of 425-cu-in. Toronado V-8 outside car is another story.

The unique chain drive, below.

How it looks with body and engine removed.

Rubber coupling appears in this photo.

them. This coupling provides up to 7.5° of wind-up for smoother power flow and freedom from shock loading. In effect it duplicates the torsional flexibility of the conventional long propeller shaft on rwd cars.

There are 4 ball-type constant-velocity universal joints of identical type and size except that the inner joints have a unique slip joint feature around their periphery which uses 30 additional balls per joint. These eliminate spline-bind when under power because shaft length changes with jounce and rebound.

The front suspension is conventional in concept but heavier and sturdier than any other GM car (because of the weight carried). The geometry is unique, however. The roll center is 1.8 in. above the ground, there is 2° of negative caster and the 11° steering pivot axis (extended) intersects the ground at a point 0.5 in. outboard of the tire tread's centerline. This and the combination of other features such as the frictionless planetary differential eliminate completely the usual front wheel drive complaints of shake when the wheels are cramped and understeer with power-on, oversteer with power off. The Olds TFD doesn't have these faults. The front springs are torsion bars and the ride rate is the highest in the industry today: 162 lb/in. Front wheel jounce travel is also a new high at no less than 5.75 in. However, we must note that the more pertinent ride frequency figure is only 72 cycles per minute, in front and with no passengers, because of the 2700-lb loading.

Torsion bars were chosen for two reasons. First, there's no room for coil springs between the upper and lower suspension arms. The coils could be mounted above the upper arm, as on the Mustang for example, but on a car of this weight the forward frame and body structure gets inordi-

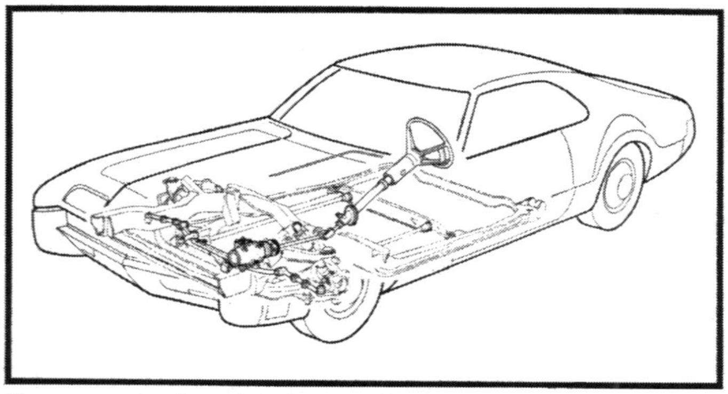

Power steering has 17.8:1 ratio with 3.4 turns lock to lock.

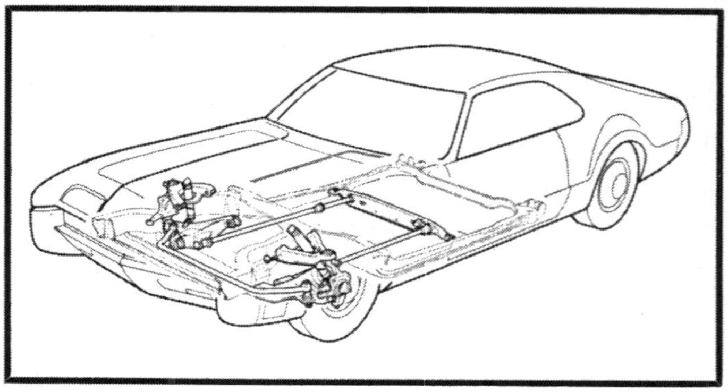

Offset torsion bar arm axes result in progressive spring rate.

Perimeter stub frame ends at front attaching point of rear springs.

Simple beam axle and single leaf springs support rear of car.

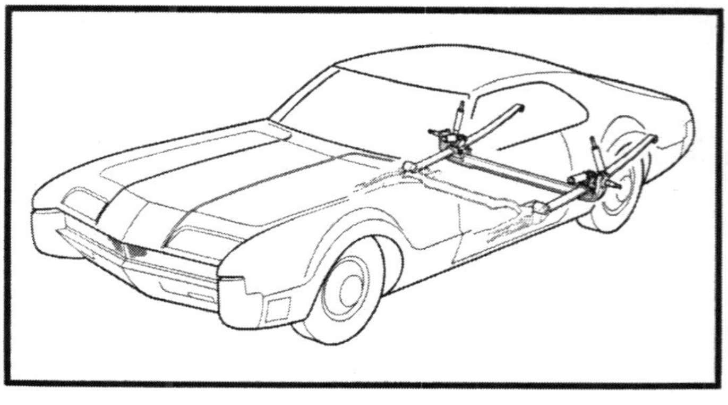

nately heavy if designed to take the load. Hence the second reason for torsion bars; the anchored ends of the bars (and concomitant loads) terminate in a special rubber cushioned crossmember under the front seat. This distributes the loading and saves structural weight. Adjustable anchors are provided for correction of front end standing height.

Power steering is used, of course; in fact it's absolutely necessary. The linkage is extraordinarily heavy and a hydraulic steering damper is used. An overall steering ratio of 17.8:1 gives 3.4 turns lock to lock and the turning circle is about average at 43 ft.

The rear suspension is very simple with a dropped beam axle and two single-leaf, semi-elliptic springs. There are, however, two shock absorbers per wheel, one vertical, one horizontal (longitudinal). The extra pair serves primarily to eliminate wheel hop and judder during heavy braking. The rear end ride rate is also an industry high at 157 lb/in. But again, as in front, the cpm drops to 96 at no load, and lower still with 6 passengers. With three in front, weight distribution does not change. Adding three more passengers (i.e. six in all) at 160 lb each lowers the rear end only 1.5 in. and the weight distribution with six is about 54/46. If each passenger contributed 44 lb of baggage, and assuming it would all fit in the trunk, the weight distribution would be 51/49 by our calculation. Thus it appears impossible ever to encounter a tail-heavy condition, or oversteer. Incidentally this situation should make the TFD an ideal tow car.

The tires are 8.85-15 and are special with the nomenclature "TFD." Normal low-profile tires are perfectly satisfactory but the TFD tires have what might be called a rain tread and the sidewalls are fractionally stiffer than standard practice. Recommended pressure is 24 psi, front and rear. Radial/belted tires, size 9.15-15, will probably be an option by the end of the year. The brakes employ large front drums (size 11 x 2.75) and the deep-dished wheels are also much heavier than usual.

This, then, is a brief summary of the most interesting new car for 1966. Oldsmobile has spent nearly seven years, millions of dollars and millions of test miles before offering it to the public. It is, by no stretch of the term, a genuine sports car. But at around $5000 it offers many special advantages that are unique and fully meritorious.

Upper and lower control arms are stampings, spindle is forged.

Two shocks per side (see text) control rear beam axle movement.

!!!AT LAST!!!

OLDSMOBILE'S FWD TORONADO

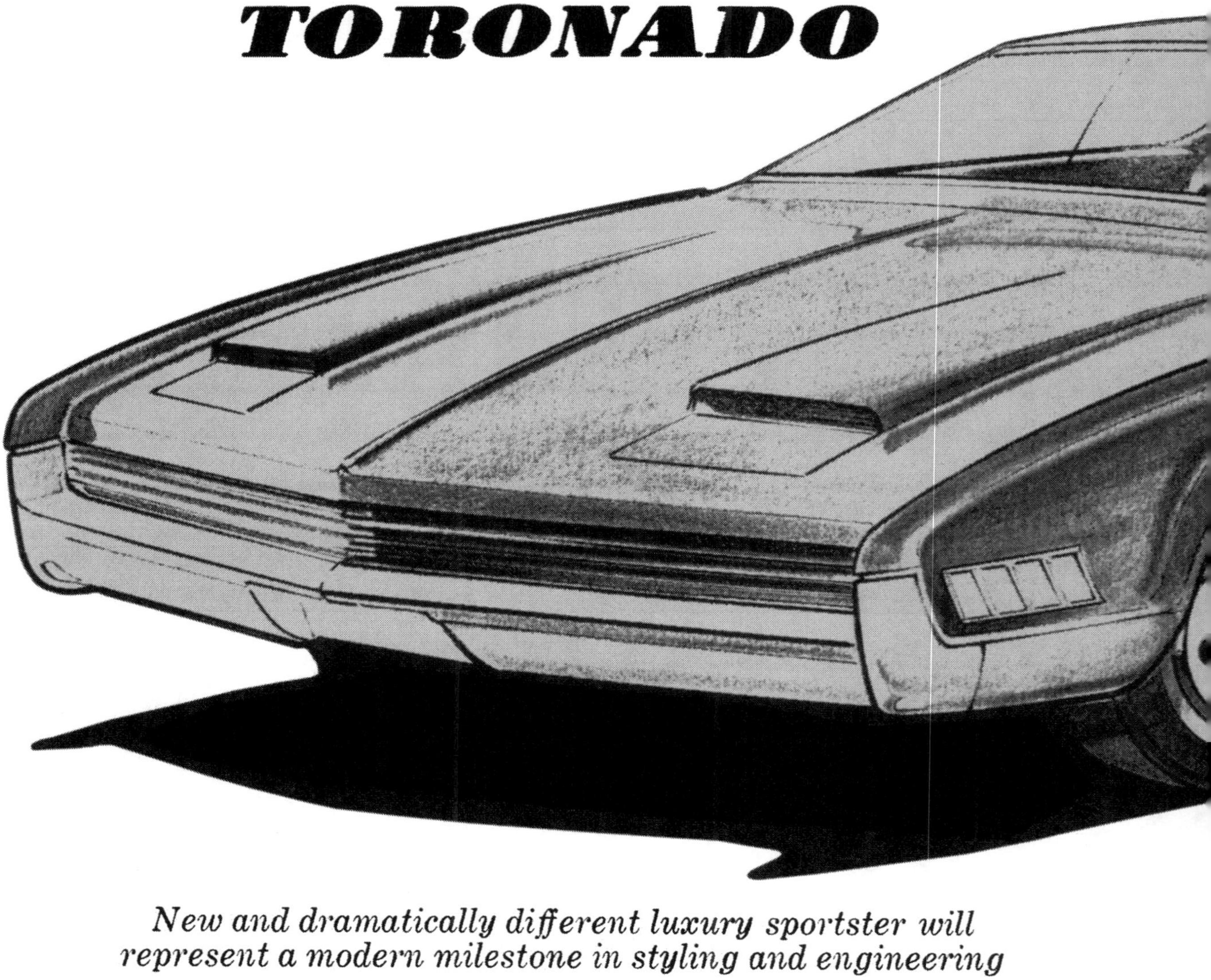

New and dramatically different luxury sportster will represent a modern milestone in styling and engineering

AFTER LONG-WINDED RUMORS and months of waiting, we've finally gotten firm news on Oldsmobile's 1966 front-drive Toronado. It'll definitely be worth waiting for. ☐ This car, as most readers know by now, will be aimed to compete with Ford's Thunderbird and the Buick Riviera – luxury cars just a shade below the high-priced three. Olds' new Toronado will share the '66 Riviera's body shell, but the styling's so different that no one can possibly mistake one for the other. About the only similarities will be retractable headlights and the absence of front vent panes. The similarity ends right there. ☐ The Toronado will be radical in everybody's book. From an engineering standpoint, its front drive will keep owners, prospective owners, mechanics, and general car bugs arguing for at least a year. The 425-inch Olds V-8 will be offset slightly to the right, with a hefty chain taking power from the flywheel to the automatic transmission. This chain runs on rubber-bushed sprockets that help absorb shock loads. The transmission, with fairly standard Turbo Hydra-Matic innards, nestles to the left of and slightly below the engine. Then, in combination with the trans, there's a special, very thin differential just in front of it. U-jointed axle shafts run off the differential, the left one naturally being shorter than the right. Finally, there's another set of universals at the outboard ends of these shafts, and the wheels are attached directly to these. Adjustable torsion bars suspend the front A-arms. ☐ Despite earlier rumors and drawings by other publications, there's no hole in the crankcase for the axle to run through. Rather, the pan has a slight indentation to accommodate it. Huge finned brakes (not discs) will give the Toronado excellent stopping power. ☐ At the rear, a short sub-frame carries a dead axle for the trailing wheels. This axle is of pressed steel, and it has an inverted hat-shaped cross section. The rear uses four shock absorbers — two pointed frontward and two vertical ones (the forward-facing ones to absorb braking forces). Then there's a single-leaf spring setup, very much like the Chevy II's. Oldsmobile engineers wasted no space with coils. In fact, the lack of a differential back there and the flat floor plan give ample room for trunk and gas tank. ☐ Inside the car, the interior will be as different as the outside. A vacuum-opened vent behind the rear seat exhausts air that comes in through the front cowl. With its completely flat floor, there's room for six all told. But the seats will look like buckets — both front and rear will have rump indentations on the outer ends, with folding arm rests between. When the arm rest's up, there's room for three people abreast, and the rest's design lets center passengers ride in complete comfort. ☐ The Cord-looking dash panel uses real, live gauges. Its speedometer revives the type Henry Ford used in his Model A, but the drum rotates vertically instead of horizontally. The door arm rests will have handles front and rear — that is, two sets, so the rear passengers can open the doors as easily as those who ride up front. Getting in and out will be no problem, because the doors open wide and high. At the curb, there's no step up and no step down — you just slide in. ☐ It may be well to pause a moment and ask why Oldsmobile decided on so unorthodox a design. They did have their reasons; it won't be just an attempt to be different. First of all, modern cars are approaching the point where they carry 60% of their weight on the front wheels. This isn't the ideal arrangement when the rear wheels have to do the driving. Basically, traction in standard cars is extremely poor, which accounts for the widespread use of limited-slip differentials, stabilizer bars, and so forth. But the new Olds will have excellent bite since that 60% concentrates right over the independently sprung driving wheels. Traction in snow, mud, in the wet, and on ice will leave most cars spinning helplessly. ☐ These benefits, of course, could come equally well with a rear-engine, rear-wheel-drive configuration. But a combination of practical and theoretical factors tipped the balance in favor of fwd — especially trunk space, engine cooling, and the adaptability of a powerplant already in production (a rear-drive plan would make for headaches in all these departments). As it stands, the chain transfer case, the all-spur-gear differential, the transaxle package, and the constant-velocity U-joint front axles were the only things

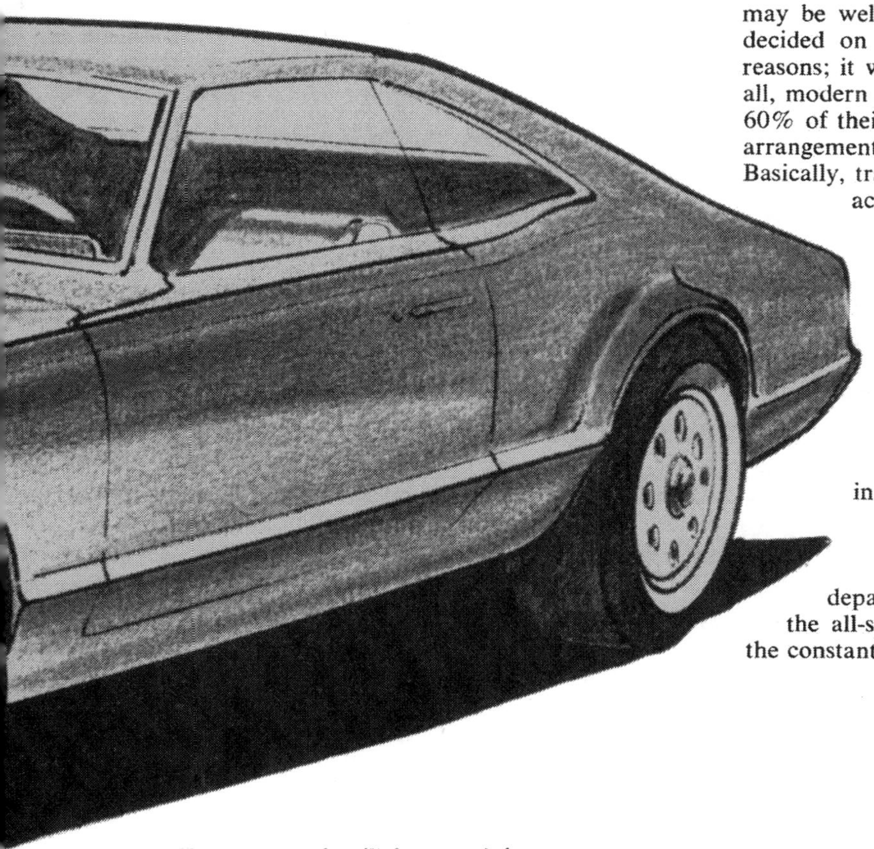

(ABOVE) *Vacuum pops headlights out of the long, sloping hood. Turning lights ride in front fenders. Toronado's height is 52 inches — same as Corvair's. Eyebrows over wheel give design a husky appearance.*

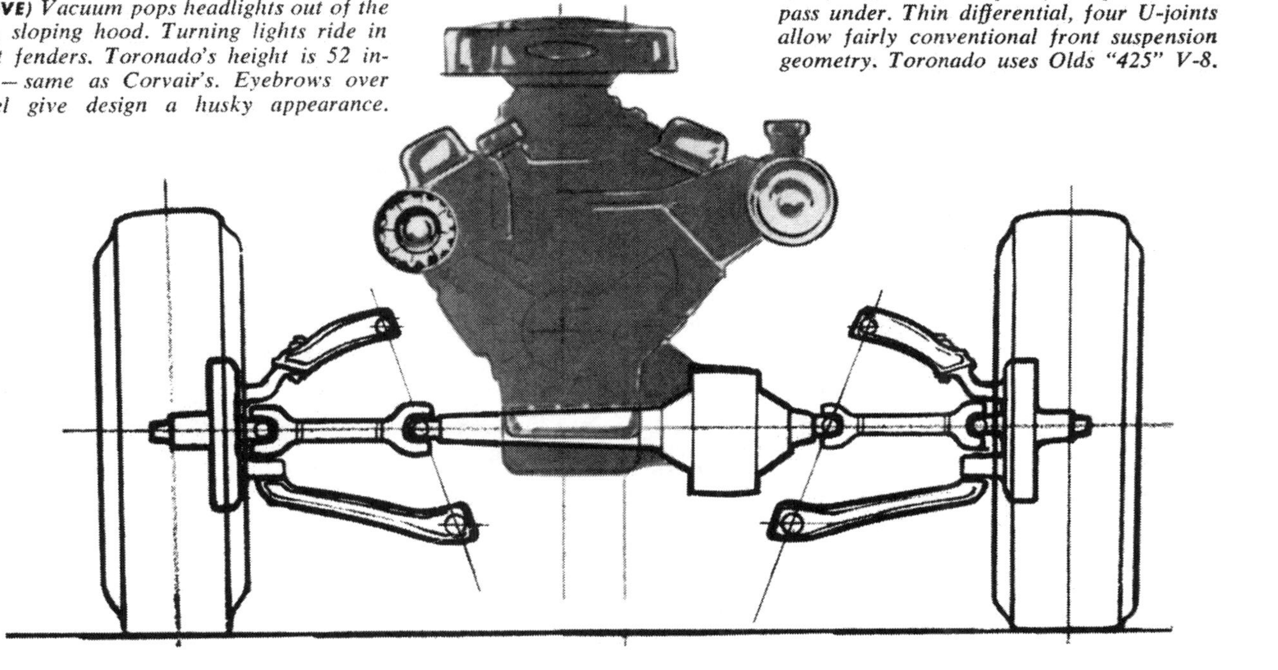

(BELOW) *Offset engine overhangs the front axles, with notch in pan for right axle to pass under. Thin differential, four U-joints allow fairly conventional front suspension geometry. Toronado uses Olds "425" V-8.*

TORONADO

Olds had to design from scratch. ☐ Then there's the matter of handling characteristics. Drivers are used to cars that understeer and, because these cars are basically stable, U.S. drivers have learned to cope with them. We expect that the new Olds will understeer less than most conventional cars with power on and plow only slightly more with it off. So there's no big change necessary in driving habits. To get good handling with a rear-engine design, you need a sophisticated, all-independent suspension system. With fwd, you don't. ☐ Although no one has used fwd in a car this large before, Olds engineers aren't sticking their necks out so much as it might seem. BMC of England has manufactured a series of highly successful fwd cars since 1960. Saab, DKW, and Citroen have used the plan even longer, and with good success. Peugeot just announced the 204, a four-door fwd car. But the question remains: Have Olds engineers solved the technical details of transmitting large amounts of torque through the steerable front axles? You can bet *they* think so, and it remains to be seen in actual tests. We should be driving the Toronado about the time you read this. ☐ Oldsmobile very obligingly showed us a five-minute movie-clip of the fwd Toronado in action. It was shot at GM's Phoenix and Milford proving grounds, and the Toronado was pitted against a standard '65 sedan. In every sequence, the fwd car outshone its cousin – it bombed up snowy grades while the other slid back down; it out-cornered the sedan on the flat; did the same when there was a rise in the middle of the corner; it stayed rock-steady at 80 mph when passing through the wind from a huge fan, while the other car veered abruptly. ☐ Part of what killed the Cord was adverse publicity. Competitors ran vicious campaigns purporting the unsafeness of fwd. Cord had to counter these with expensive advertising, but even so, it didn't help. We've seen somewhat the same thing recently when competing compacts downgraded the early Corvair. Remember the TV commercials showing the instability of an arrow shot backward? Oldsmobile, we feel, expects at least some of the same when they formally unveil the Toronado. From the brief look of the Toronado as we've seen it, Olds has nothing to worry about. But the real test will come when someone on the outside has driven the car, and that's a pleasure we hope to report within the next two or three months.

ILLUSTRATIONS BY TOM DANIEL

VERY SHORT REAR DECK MAKES THIS A TRUE FASTBACK. FLAT FLOOR GIVES PLENTY OF TRUNK ROOM. VENT UNDER WINDOW EXHAUSTS AIR.

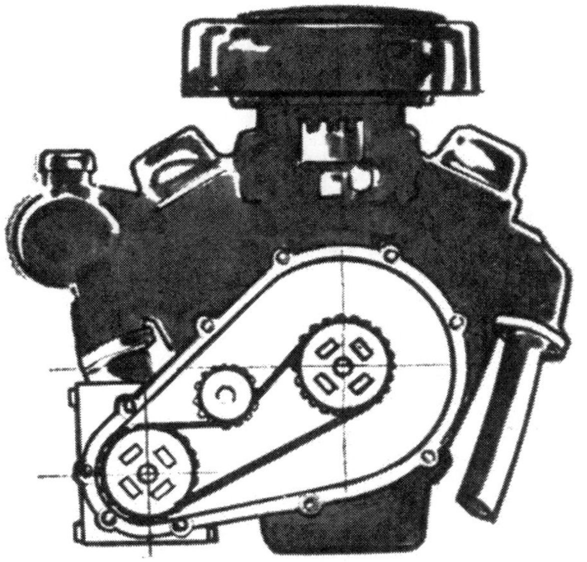

Chain off rear of engine drives transmission, which nestles below block. Thin differential (not shown) stands in front of gearbox.

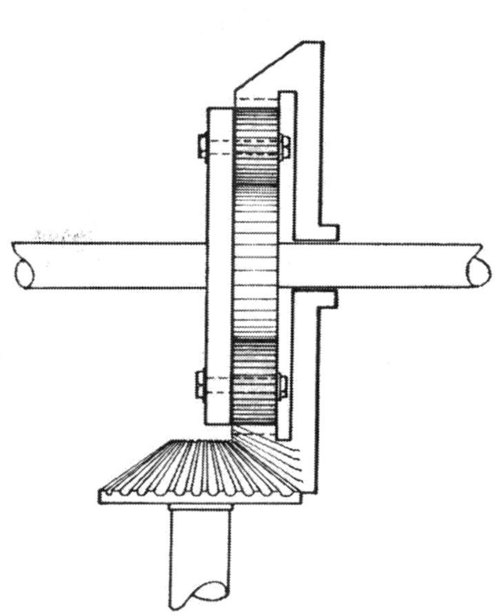

Schematic of all-spur, planetary-gear differential shows it's not much wider than pinion, squeezes between V-8 and U-joints.

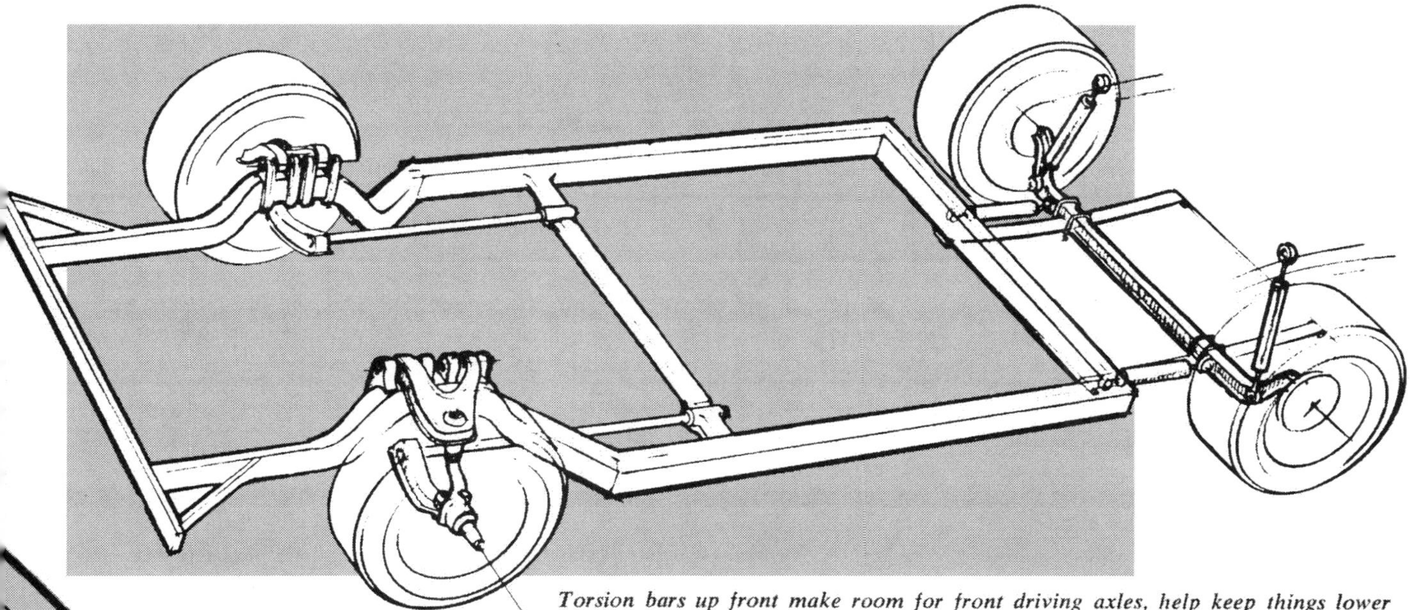

Torsion bars up front make room for front driving axles, help keep things lower than coils would. Perimeter frame uses box-section girders with smaller subframe behind. Single-leaf springs mount four shocks, with two pointing forward.

(RIGHT) *Wide interior, flat floor make room for six people. Dash uses working gauges, drum speedometer. Double door handles allow rear passengers to let themselves exit.*

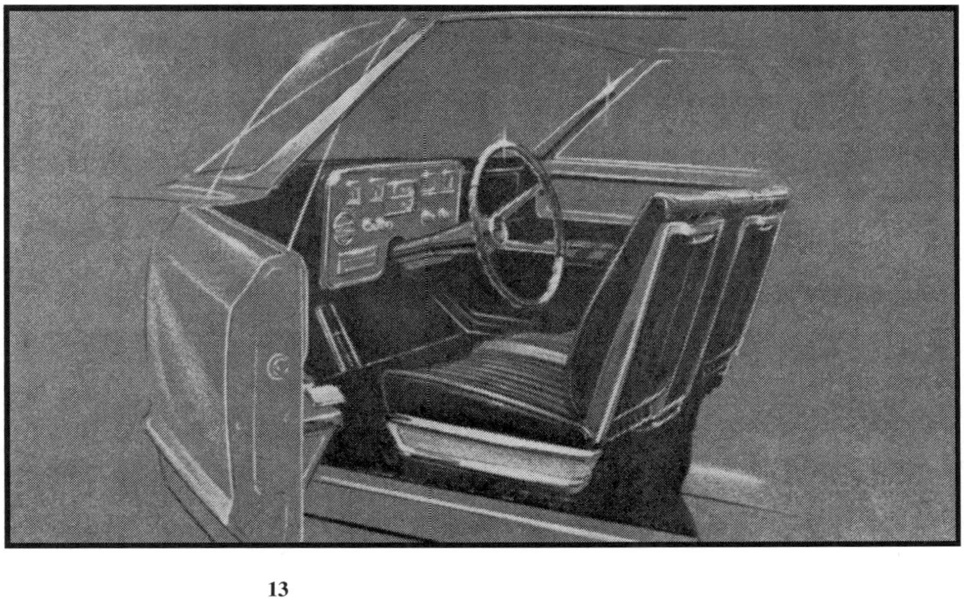

TORONADO vs. RIVIERA

AN ON-THE-ROAD COMPARISON

At first look, the only logical reason for buying an Oldsmobile Toronado over the Buick Riviera might be to obtain the mechanical novelty of front-wheel drive. Size, styling, passenger accommodation and performance, at least between Riviera and Toronado, are nearly on a par. But, where the Riviera retains tried-and-true, traditional front engine/rear-wheel drive, the Toronado offers a new front engine/front-drive arrangement. It's a powerful selling point, but is it a justifiable one?

Car Life sampled two early-production versions of the Toronado and Riviera and concluded that the Riviera is the better planned and finished car and that the Toronado is more roadable and stable under varied conditions.

In this latter context, the Toronado is outstanding. It has styling distinctive enough to make it stand out in any parking lot, and that styling shouts, "This is a front-wheel drive!" In a time when the outpouring of U.S. automotive design not only looks very much similar but is virtually identical in mechanical specification, the Toronado at least offers its buyer relief from the endless similitude.

The Riviera, too, has a distinctive styling and CL's reviewers found it the more refined and tasteful. But the Riviera has a drive-train just like that of every other Buick produced, so can offer no special appeal in this area.

Different people prefer different things. Good taste is good taste, no matter who likes or dislikes a design. The Riviera is a good example of good taste and good design. The Toronado is reasonably good design, but because of less refinement of its lines is not necessarily an example of good taste; it isn't gross enough in trim, proportion or finish to be in bad taste, it just doesn't come off as gracefully curved and pleasingly proportioned as does the Riviera. Side-by-side, or nose-to-nose, comparison quickly reveals the differences. The Riviera proclaims by purity of line and understatement that its occupants have enough confidence in their tastes to eschew ostentation.

Interior fittings tell the biggest story. The Riviera has all the elegance and attention to detail one must expect in a $5000 car. The Toronado appears to have been short-changed in this area in order to offset the expense of the much-costlier drive-train.

The Toronado with its flat floor develops one distinct advantage over the Riviera. With the standard bench front seat, the Toronado easily accommodates six adults. With the same seat, the Buick also seats six, but the two in the middle have to be either short-legged or uncomfortable. Otherwise, seating, knee-room, head-room and leg-room dimensions are virtually identical.

Straight-line accelerations are roughly comparable, unless weather or road-surface conditions are added to the consideration. Then the Toronado's greatest single advantage is immediately apparent. Wheelspin on takeoff is virtually impossible, even when Toro's front wheels are inches-deep in water. Traction on muddy, rain-slick streets is phenomenal —every bit as good as Olds claims it to be. On the other hand, the Riviera's rear-wheel drive skitters and slips all over the place when the throttle is injudiciously applied on even dusty pavement. The reason is obvious: Toronado has 61% of its test weight on its drive wheels where the Riviera has only 45%.

Over-the-road handling is the final major consideration and here again the Oldsmobile comes out ahead. Though a driver needs some experimentation before he can get the most out of the car, he will find the Toronado drives much like a normal car. On the other hand, the Riviera handles like a normal car in all situations; straight-line driving is comfortable and non-traumatic, curves taken too fast result in the car plowing off the road nose-first. The Toronado, has the great understeering nose-plow, too, but something can be done about it. When the driver finds he has entered the turn too fast and is being led head-first off the outside of the curve, he can slant his front wheels toward the inside of the curve, back off and then stand on the throttle and let the front wheels pull the car on around the turn. But, on straight, or mildly curving expressway, few non-enlightened drivers could tell the difference between the front and rear-wheel drives.

The Riviera scores mightily over the Toronado in the braking tests, achieving deceleration rates of 22 and 21 ft./sec./sec. in the first two all-on stops from 80 mph. The best the Toronado could do was 18 and 12 ft./sec./sec. and that was accompanied by rapidly building brake fade. Riviera went to five consecutive stops from 80 before fade made the brakes temporarily unreliable. Here, the hefty forward weight bias works against the Toronado.

The strong points for the Riviera, then, are quality of finish and esthetic appeal and good brakes. For the Toronado, advantages are a soundly engineered drive system which produces outstanding traction for both handling and adverse road conditions, and better utilization of interior space.

Either car should prove reliable and durable in the hands of the owner-driver. Both have a mechanical quality far above many cars in the same price range. But, the choice still comes down to whether or not the buyer with $5000 to spend wants the conversationally prestigious attributes of front-wheel drive over the esthetic appeal of the rear-wheel drive model. ■

CAR LIFE
ROAD TEST

NOT SINCE THE days when a British sports car had the power to gather a knot of awed people wherever it stopped has one automobile proved such a crowd-catcher. Never has any automobile elicited so much comment, drawn so many bystander stares. Heretofore it was easy to believe much of the American public is blasé about cars, regarding automobiles simply as a means of traveling from one place to another with speed and in comfort. Hence the reaction of a cross-section of the public to this car could be expected

OLDSMOBILE TORONADO

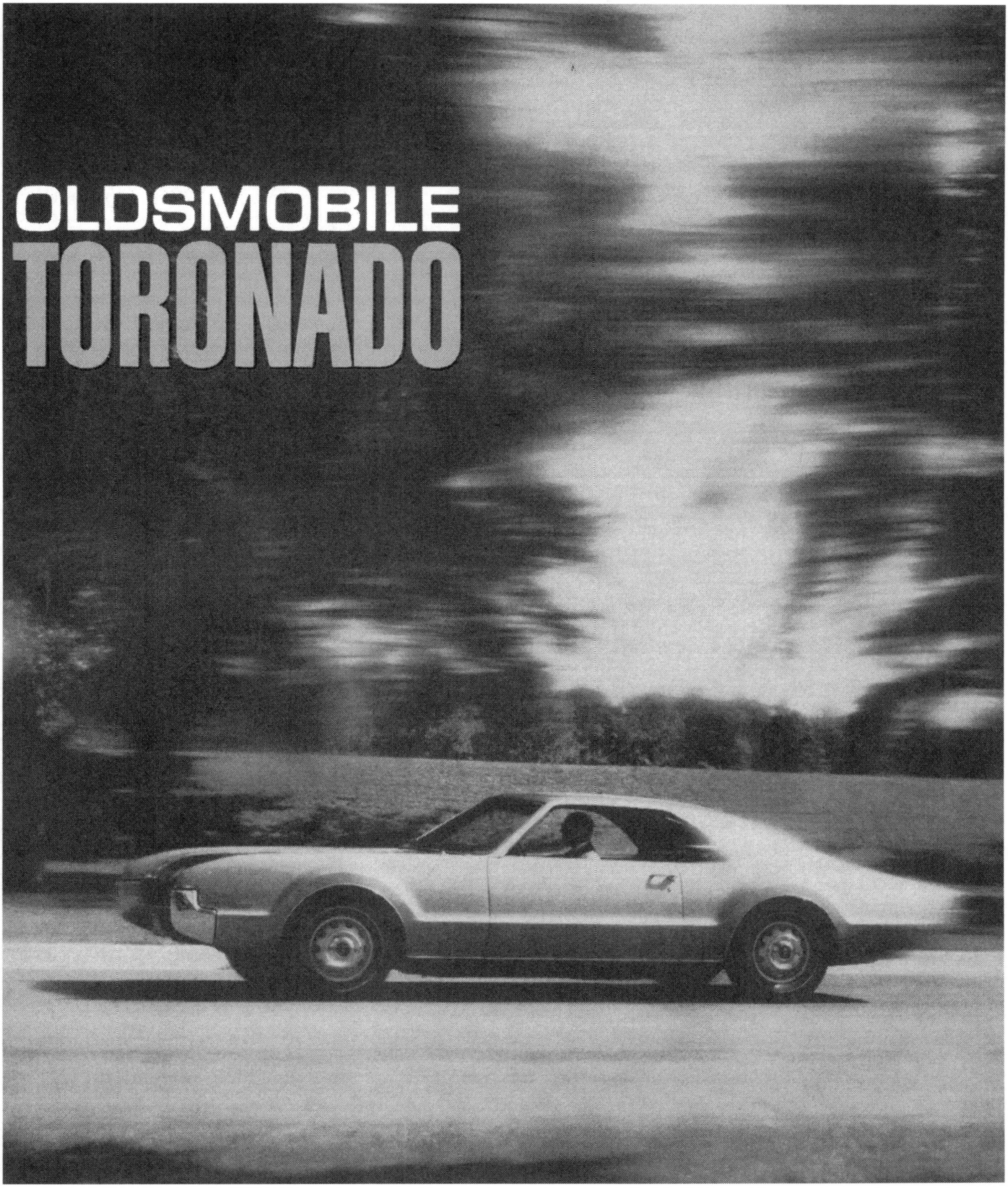

TORONADO

to be one of indifference. Such was not the case. Everywhere the car was taken during *Car Life*'s test period, everyone encountered expressed an opinion—positive or negative—or else asked a question. Everyone had something to say.

Elderly Lady (slightly deaf): 'What kind of car is it?"

An Oldsmobile.

"Ninety-Eight?"

No, Toronado.

"Tornado?"

No, Toronado.

"Coronado?"

No, *Toronado*.

"Why?"

Why indeed? For a big, bold fastback *gran turismo* automobile with an exceedingly strong engine, and treated with infinite care in drive-train engineering, the name Toronado has a somewhat pusillanimous ring. Come right out and call 425 cu. in. a Tornado. Call it a Coronado, for that is a luxurious Southern California resort community named after an intrepid Spanish explorer, hence quite a fit place for visiting in a powerful 6-passenger modern American touring car. Toronado seems a name, like a camel, designed by a committee.

Jackhammer Operator (at water main repair diggings, midtown): "Is this one of those front-wheel-drive rigs?"

Yes. The Toronado is a fwd automobile, the first built in the United States since the 810/812 Cord, production of which ceased in 1937. Design engineering, not to be confused with production engineering, that produced the Toronado's fwd system is of such merit the car was named for *Car Life*'s 1966 Award for Engineering Excellence. The Toronado's departure from the American standard of front engine/rear drive incorporates the standard Oldsmobile torque converter in its normal position at the rear of the V-8 engine block. From that point, a chain drive carries power to the standard Oldsmobile 3-speed Turbo Hydra-Matic transmission, rotated 180° to a position at the left side of the engine. Power is then transferred to a specially designed, very slender planetary differential. Torque is delivered to the front-drive wheels from the differential by half-axles of unequal length.

Boxout Boy (at neighborhood supermarket): "Does it handle any differently than a regular car?"

That boxout boy wasn't the first or the last interested bystander to voice that question in regard to the Toronado's roadability. Does the Toronado handle differently? The answer must be a qualified yes and no. The qualifications stem from road surfaces and degrees of curving road. On a smooth freeway at 65 mph or thereabouts, the handling of the Toronado cannot be distinguished from that of a rear-wheel-driven automobile of like weight and size, and fitted with power assists similar to those of the Toronado. Hence the answer could be: "No, the Toronado's handling is no different than a regular car." But, put *el Toro* on a skein of winding roadway, narrow and semi-mountainous. Here the effect of the fwd system may be assessed.

Understeer, the tendency for an automobile to remain on an established course despite desired change in wheel angle cranked in through the steering mechanism, is the Toronado's major characteristic.

Electronics Engineer (who takes all major automotive magazines): "It understeers pretty bad, I guess."

Yes, the Toronado understeers, but not badly, if the driver learns and exercises a revised set of cornering techniques to accommodate the fwd system. The Toronado's understeer, of course, is generated by the forward weight of the engine, drive-train components, battery and assorted accessory pumps, hoses, pulleys and belts. Only 40% of the Toronado's total weight is behind the firewall. Handling this frontward weight in tight bends, however, is simplicity. Hard into the corner, when it seems the Toronado's massive front end will slew toward the outward arc of the bend, all that is required is a momentary letup on the accelerator pedal, then a re-application of power to re-establish the track through the curve. The Toronado, in effect, has the capability to pull itself through bends with power applied to its front driving wheels. The feel of taking a bend at good speed in the Toronado is different than sensations experienced in cornering a front engined/rear driven car of like stature. The Toronado driver can't break the rear end loose for a drift with application of throttle—because the Toronado's free-wheeling rear end simply follows where the front drive wheels lead it. Therefore, to avoid an untoward shunt into a barbed wire fence, the trick is to make the front wheels do an accurate job of leadership with judicious see-sawing of the accelerator pedal. This requires practice.

Test drivers noted that when new to the Toronado, they found it impossible to judge where the right front wheel was located on the pavement. The proportions of the Toro are so different from standard that the car requires a great deal of getting used to before the driver can become confident that he is not creating traffic hazards by unknowingly locating parts of the car in traffic lanes other than his own.

Blue-Eyed Girl Child (age 5, on way to kindergarten): "It's smooth."

One must agree that the Toronado is smooth, offers a boulevard ride—on boulevards. At high speeds on straight expressways the passenger compartment seems to float, bobbing a bit like a balloon being towed in mildly turbulent air by an elephant. On rural roadways, sometimes seamed and humped, speed will cause the light hindmost end of the Toronado to flap like a sheetmetal shirttail. Rough pavement does not destroy the positive track of the front drive wheels, but the suspension system for the heavy drive components works hard and somewhat ponderously in the attempt to soften the ride. Altogether, the ride offered by the Toronado is smooth—like Rocky Road ice cream, which has a few lumps here and there.

THE MAJOR exception to the smoothness observation occurs when one of the Toro's drive wheels drops into a chuckhole. This condition produces a thundering rebound, but not loss of control, due to the automated regulation of power steering, although the

FLAT FLOORS please passengers, but 11-in. drum brakes show alarming fade characteristics.

action is somewhat heavier than one would anticipate for a full power-assisted system.

Panhandler (near shoestore, seeking dimes and quarters): "How does she go?"

The simple answer was, "Very well." Though the ragged, destitute one didn't ask, there are a number of reasons why the Toronado goes well. The Toro's 90° V-8 engine is oversquared at 4.125 in. to a stroke of 3.975 in. With a 10.5:1 compression ratio, the 425 cu. in. develop 385 bhp at 4800 rpm. Translated into auto operation, this means the big, relatively lazy, slow-turning engine will produce an effortless cruise at 80 mph and will, without undue prodding, exceed all sorts of speed limits. Frankly, the 115 mph and more of which the Toronado is capable, is far too fast for intelligent driving.

At the red light raceway, the Toronado can easily be beaten off the line. Quarter-mile e.t.s of 17-18 sec., nonetheless, are somewhat phenomenal when considered in the light of the Toronado's total weight: Something in excess of 5000 lb. as tested. Drag racing isn't the Toronado's forte. The effortless power surge, a hearty reserve which may be tapped at 65 mph and beyond, is the shining performance feature.

If the Toronado has very adequate go, it also has totally inadequate stop. The engineering which resulted in the Toronado's drive system should have extended to the car's braking system. Drums of 11-in. diameter and shoes of 2.75 in. width in front and 2 in. width in the rear provide a total swept area of 328.2 sq. in. This is insufficient for a car whose gross weight is in excess of 5400 lb. when fully passengered, fueled and conservatively cargoed. In two intentional panic stops from 80 mph, the Toronado brakes demonstrated alarmingly unacceptable fade characteristics, that is, lockup which induced noisy slides and anything but straightline deceleration. A totally unintentional panic stop in a freeway situation left one *Car Life* driver with trembling hands, icy perspiration and a total lack of desire to drive the Toronado ever again . . . unless the car is given discs or a front discs/rear drums, with a limiting device (*à la* Thunderbird, Continental and Ford 7-Litre) to prevent rear-wheel lockup. The present drum system, without such limiting, allows the wheels under the light rear end to halt completely for sustained, squealing, squirming slides. All of this is in the light of the statement by Oldsmobile engineers: "The Toronado does not need disc brakes." *Car Life* disagrees.

College Professor (an MG-TD owner, as companion on a cross-town drive): "Man, the leg room. I feel like Alice in Wonderland—growing ever smaller."

Leg room for average-sized persons is adequate in the Toronado. Though the good professor, at 5 ft., 9 in., found more than a sufficiency of leg room, *Car Life*'s 6 ft., 3 in., staffer did not—but his is an unusual long-legged case. His view was that an overall length of 211 in. on a 119-in. wheelbase could well provide more space for passengers of lengthy underpinning.

Width is no problem in the Toronado interior. Hip room is 62.2 in. in front, and a cozy-for-three 55.6 in. in the rear.

Headroom, slightly more than 37 in. front and rear, is something else again. Though adequate for front seat passengers, the fastback roofline, which slopes rearward to a chopped-off stern, curves close to the head tops of rear seat passengers. One assessment is that the Toronado could have been made a much roomier car lengthwise—but it wouldn't have been a fastback.

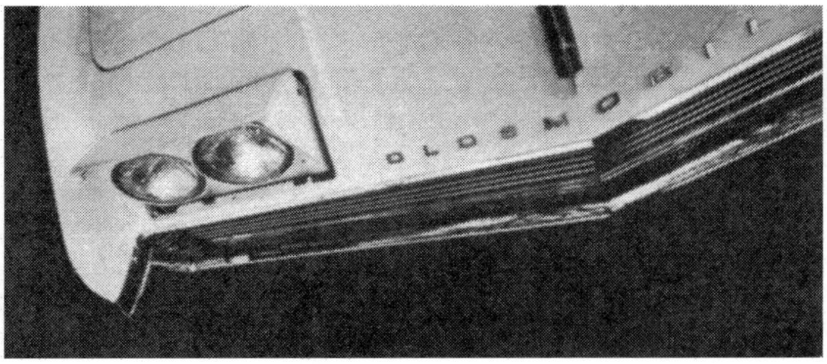

CHAN BUSH PHOTOS

THE TORO features vacuum actuated headlamp units and a wide, solid bar of tail lamp illumination.

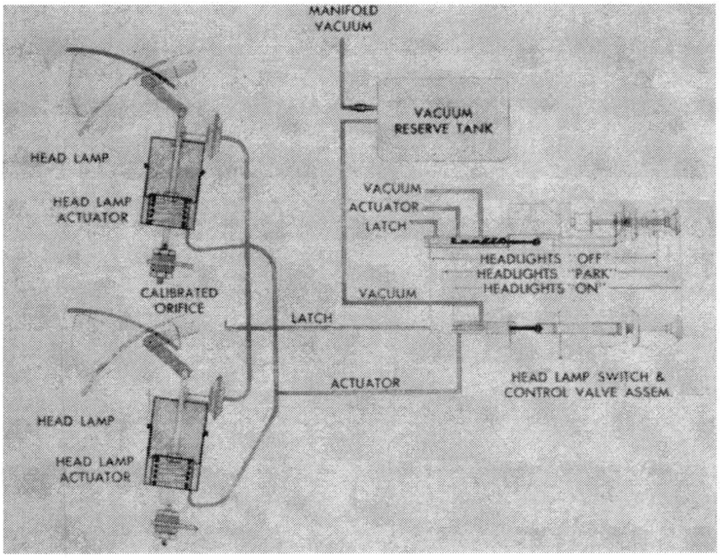

TORONADO

Because all mechanicals are forward, no driveshaft tunnel arches its back into *el Toro*'s floorspace. All six passengers are given space—pool-table flat—to place the soles of their shoes. There are no overly updrawn knees for center seat passengers in the Toronado.

The expanse of flatness—like Kansas—extends into the luggage compartment, where the spare is buried at the back of the rear seat, which requires all cargo to be removed for a wheel change.

Housewife (from next door, wiping hands on apron): "My, what a lovely car!"

The exterior lines of the Toronado are impressive. The car has massive size, a brute functional look Toronado features at a glance are sharp angling front fender lines, large (8.85-15) tires and steel spider stamped wheels (15 x 6JK), a long, sweeping expanse of hood and sheetmetal convolutions which emphasize the muscular look at each wheel. Indeed, from a distance of no less than 25 ft., the Toronado is an attractive car. Closer inspection, however, brings to view a few irregularities of panel fit, but none that are not to the current standard of Detroit workmanship.

Junior High School Student (having skateboarded dangerously across the flow of residential traffic for a closer look): "Neat!"

Hardly. The test Toro's paint, something approaching a rich, royal Satsuma plum, was applied in good fashion, but beneath the paint, in the passenger compartment the flaws of manufacture (production engineering, as distinguished from design engineering) become apparent. Upholstery, in matching shades of vinyl and nylon, was well fitted, but appeared inappropriate for a GT sort of automobile. Full leather, or at least full leather-like vinyl, would be more in keeping with the Toronado's outward aura of high speed touring. The impression was one of seeing lace curtains and shelves of china knick-knacks in a clean, well-fitted machine shop.

Pillar and window moldings fitted well at some points and did not at others. The large, bin-like glovebox was sticky and required two hands to open—one on the release catch button, the other prying at the bin cover.

Door hardware, specifically a sort of gum wrapper catchall underneath the door latch lever, set up sympathetic vibrations for a distressing buzz. This particular sound, like so many others in the car, is the result of an indefensible lack of craftsmanship in assembly. Evidence of shoddy workmanship appeared suddenly when a piece fell off the Toronado onto a test driver's foot. This was later determined to be a fairing, a sort of scuff shield, which had received cavalier treatment during factory installation. This sort of thing should not happen.

Hand-crank window lifts on the Toronado? Yes. These were definitely out of place on this Oldsmobile of all Oldsmobiles. Power windows are offered only as an option. Also out of character is the manually adjusted front seat, with only backward/forward adjustments possible.

The left-hand window lift mechanism on the *Car Life* test car tended to get out of whack when the window was rolled to its most downward position. An Olds dealership mechanic fiddled the gearing back into operation in the manner of a chiropractor, putting things to right from the outside without resorting to major surgery. His comment was, "This happens a lot.

1966 OLDSMOBILE

TORONADO HARDTOP COUPE

DIMENSIONS

Wheelbase, in.	119.0
Track, f/r, in.	63.5/63.0
Overall length, in.	211.0
width	78.5
height	52.8
Front seat hip room, in.	62.2
shoulder room	58.8
headroom	37.9
pedal-seatback, max.	48.0
Rear seat hip room, in.	55.6
shoulder room	57.8
leg room	35.5
head room	37.5
Door opening width, in.	48.3
Floor to ground height, in.	12.3
Ground clearance, in.	5.0

PRICES

List, fob factory	$4812
Equipped as tested	5730

Options included: Glare-proof rearview mirror, courtesy and warning door lamps, visor vanity mirror, luggage compartment lamp, underhood lamp, am/fm radio, rear speaker, wsw tires, chromed spider wheels and trim rims, tinted glass, air cond.

CAPACITIES

No. of passengers	6
Luggage space, cu. ft.	14.5
Fuel tank, gal.	24.0
Crankcase, qt.	5.0
Transmission/diff., pt.	24.0/n.a.
Radiator coolant, qt.	18.0

CHASSIS/SUSPENSION

Frame type: Boxed perimeter/integral

Front suspension type: Independent s.l.a. ball joints, torsion bar springs, tubular shock absorbers, link-type stabilizer.

ride rate at wheel, lb./in.	162
anti-roll bar dia., in.	1.0

Rear sɑspension type: Single leaf springs, beam axle, horizontal and vertical shock absorbers.

ride rate at wheel, lb./in.	157

Steering system: Power-assisted recirculating ball nut; parallelogram linkage with hydraulic damper.

gear ratio	17.5:1
overall ratio	17.8:1
turns, lock to lock	3.4
turning circle, ft. curb-curb	43.0
Curb weight, lb.	4660
Test weight	4970
Weight distribution, % f/r	61/39

BRAKES

Type: Single line hydraulic with self-adjusting duo-servo shoes in finned cast-iron drums.

Front drum, dia. x width, in.	11 x 2.75
Rear drum, dia. x width	11 x 2.00
total swept area, sq. in.	328.2
Power assist	integral, vac. booster
line psi @ 100 lb. pedal	1125

WHEELS/TIRES

Wheel size	15 x 6JK
bolt no./circle dia., in.	5/5.00
Tire make, brand	Firestone Deluxe Champion
size	8.85-15
recommended inflation, psi	26
capacity rating, total lb.	n.a.

ENGINE

Type, no. cyl.	ohv, V-8
Bore x stroke, in.	4.125 x 3.975
Displacement, cu. in.	425
Compression ratio	10.5:1
Rated bhp @ rpm	385 @ 4800
equivalent mph	124
Rated torque @ rpm	475 @ 3200
equivalent mph	82
Carburetion	1 x 4
barrel dia., pri./sec.	1.375/2.25

Valve operation: Hydraulic lifters, pushrods and rockers.

valve dia., int./exh.	2.067/1.629
lift, int./exh.	0.431/0.433
timing, deg.	21-77, 71-31
duration, int./exh.	278/282
opening overlap	52

Exhaust system: Dual, transverse muffler and tailpipe resonators.

pipe dia., exh./tail	n.a.
Lubrication pump type	gear
normal pres. @ rpm.	30-45 @ 1900
Electrical supply	alternator
ampere rating	42
Battery, plates/amp. rating	78/73

DRIVE-TRAIN

Transmission type: Torque converter automatic with chain driven planetary gear transmission.

Gear ratio 4th ( ) overall	
3rd (1.00)	3.21
2nd (1.48)	4.65
1st (2.48)	7.95
1st x t.c. stall (2.20)	17.5
synchronous meshing	planetary
Shift lever location	column

Differential type: Front-wheel drive planetary

axle ratio	3.21:1

This curved side glass sticks in the rubber molding."

The Toronado's revolving drum speedometer is exceptionally readable, though the speedometer drive cable developed a fluttering paradiddle which came and went without any particular pattern.

A PHONY knurled wheel, to the right of the heater/air conditioner pushbutton panel (a match for radio tuning controls on the opposite side of the dash) seems to carry symmetry a bit far. The "nothing control," as it was dubbed by test crewmen who had been suckered into attempting to see what the knurled wheel regulated, is designed to take up space when a speed control device is not installed.

Perfectly aligned and sized indicator lights, located above four small gauges (three are operational pointers, one is a warning light), are just too precious. Two are for turn signals, one is a parking brake warning and the remaining one is the high beam indicator. In the latter, only a tiny dot in the center of the oblong lens comes alight, rather than the entire rectangle as is the case with the other lights. So much for symmetry.

Such things may seem trivial when taken singly, but to *Car Life* staff members the sum total of these manufacture and production engineering deficiencies indicates that Oldsmobile has made up in interior, trim and fittings the amounts of money lavished underhood, especially on the drive-train. The effort was apparently to keep the Toronado's total price down.

Gas Pump Jockey (on being informed that the Toronado's price is upward from $4600): "This sure ain't the car for a guy like me—a poor man."

He's probably right. Upward from $4600 means state taxes, licensing and insurance may be added at that point. Dealer servicing and extras could add a bit more. The Toronado option catalog includes a 6-way power bench seat, tilt/telescope steering wheel, a reclining front passenger seat, headrests, an am/fm radio, rear seat speaker, power antenna, air conditioning, rear window defroster, speed warning, speed control, air injection reactor (mandatory in California), luggage compartment lamp, underhood lamp and cornering lamps among other things. Though the total option slate might be somewhat spare in comparison to offerings with other makes of cars, addition of very much equipment could push the cost of the Toronado well over $5500.

Barber (trimming sideburns): "How many miles per gallon does it get?"

Fuel economy is a consideration not restricted to the "poor man." The only answer to the question lies in *Car Life*'s test figures. Conservative driving produced 11.3 mpg. Hard and high speed use resulted in a fuel reading of 10.2 mpg. These are neither the worst nor the best fuel consumption figures tallied among American V-8 engined automobiles, but they indicate $3.50 gasoline cost per 100 miles.

Physicist (who drives a Mercedes-Benz 180 and has purchased a Ford Falcon station wagon for his wife): "Who'd buy a car like that anyway?"

A guess would be the purchaser of the Toronado is one who desires first-class transportation, something that isn't exactly like every other car on the road—something new. This person may be an adventurer who is too portly for a Porsche, not junior enough for a Jaguar, not middle-income enough for a Mustang, or not *bon vivant* enough for a Bonneville convertible. The Toronado may not appeal to youth.

Delicatessen Clerk (a daytime college student in chemistry and a Corvette owner): "When do you think Chevrolet will bring out the Mako Shark?" ■

CAR LIFE ROAD TEST

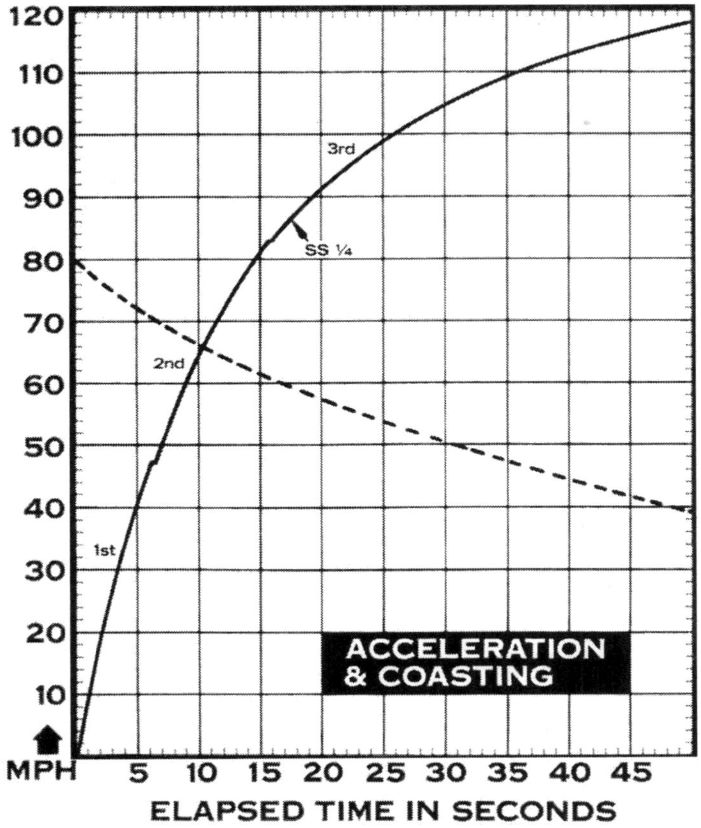

CALCULATED DATA

Lb./bhp (test weight)	12.9
Cu. ft./ton mile	114.5
Mph/1000 rpm (top gear)	25.7
Engine revs/mile (60 mph)	2330
Piston travel, ft./mile	1540
Car Life wear index	36.0
Frontal area, sq. ft.	23.0
Box volume, cu. ft.	506

SPEEDOMETER ERROR

30 mph, actual	29.3
40 mph	38.7
50 mph	47.9
60 mph	57.7
70 mph	66.7
80 mph	78.5
90 mph	88.2

MAINTENANCE INTERVALS

Oil change, engine, miles	6000
transmission/diff.	24,000/none
Oil filter change	6000
Air cleaner service, mi.	24,000
Chassis lubrication	6000
Wheelbearing re-packing	as req.
Universal joint service	none
Coolant change, mo.	24

TUNE-UP DATA

Spark plugs	AC 44S
gap, in.	0.030
Spark setting, deg./idle rpm	7.5/850
cent. max. advance, deg./rpm	20/4200
vac. max. adv., deg./in. Hg.	20/20
Breaker gap, in.	0.016
cam dwell angle	28-32
arm tension, oz.	19-32
Tappet clearance, int./exh.	0/0
Fuel pump pressure, psi	7.75
Radiator cap relief press., psi	15

PERFORMANCE

Top speed (5200), mph	135
Shifts (rpm) @ mph	
3rd to 4th ( )	
2nd to 3rd (4800)	83
1st to 2nd (4500)	47

ACCELERATION

0-30 mph, sec.	3.2
0-40 mph	4.9
0-50 mph	6.8
0-60 mph	8.9
0-70 mph	11.5
0-80 mph	14.5
0-90 mph	19.0
0-100 mph	26.0
Standing 1/4-mile, sec.	17.8
speed at end, mph	86
Passing, 30-70 mph, sec.	8.8

BRAKING

(Maximum deceleration rate achieved from 80 mph)

1st stop, ft./sec./sec	18
fade evident?	yes
2nd stop, ft./sec./sec.	12
fade evident?	yes

FUEL CONSUMPTION

Test conditions, mpg	10.2
Normal conditions, mpg	10-13
Cruising range, miles	240-312

GRADABILITY

4th, % grade @ mph	
3rd	10 @ 67
2nd	17 @ 54
1st	29 @ 36

DRAG FACTOR

Total drag @ 60 mph, lb.	210

Talk about rare breeds—you're looking at one: new Toronado by Oldsmobile. One-of-a-kind handling with front wheel drive to put the action where the traction is. One-of-a-kind performance from a 385-hp Toronado Rocket V-8 and a specially tailored Turbo Hydra-Matic. One-of-a-kind comfort (for six, mind you) a la flat floors and draft-free ventilation. One-of-a-kind convenience from such niceties as concealed headlamps and flat-floored trunk. And (rather obviously) one-of-a-kind style too! Why not stop in to see your one-of-a-kind Olds Dealer! LOOK TO OLDS FOR THE NEW!

VITAL STATISTICS

Wheelbase	**119"**
Overall length	**211"**
Overall width	**78.5"**
Overall height	**52.8"**
Curb weight	**4496 lbs.**

Standard Equipment: Turbo Hydra-Matic • Power Steering • Power Brakes • Deluxe Steering Wheel • Foam-Padded Front Seat • Parking Brake Lamp • Carpeting • Clock • Heater-Defroster • Front and Rear Seat Belts • Padded Instrument Panel • Padded Sun Visors • Glare-reducing 2-speed Wipers • Washers • Outside Mirror • Directional Signals • Back-up Lamps

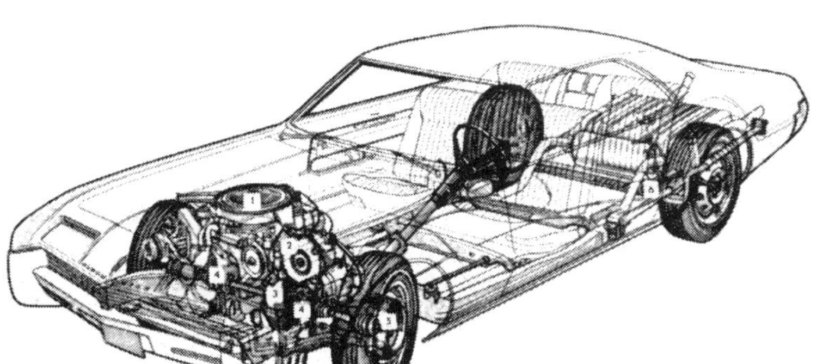

What makes Toronado tick?

1 425-cube Toronado Rocket V-8 with Quadrajet Carb delivers 385 hp, 475 lb.-ft. torque, 10.5-to-1 ratio. 2 Newly designed Turbo Hydra-Matic features variable-vane torque converter. 3 All-new low-friction planetary differential is exclusive Olds design. 4 Front drive shafts are equipped with twin, constant-velocity universal joints for smooth power flow. 5 Burly wheels with 8.85 x 15-inch tires, pull (not push) Toronado through turns and over rugged terrain. 6 Rear leaf spring, two vertical and two horizontal rear shock absorbers provide firm yet comfortable ride.

OLDSMOBILE TORONADO

Will it be only the first of many American fwd cars?

NO AMERICAN CAR has caused such a flurry of technical interest as the Toronado has since that other extreme of American car design, the Corvair. Having devoted many pages to analysis of the engineering that went into the development of this new car (R & T November 1965), we now turn to an examination of the finished product from the standpoint of the daily user.

To those who had hoped the Toronado would be a departure in American car design, the car must be a disappointment. The practice of grouping the entire power train at one end of the car has become common in Europe over the past few years, with the front engine/front drive concept gaining rapidly on the rear/rear layout. And there's a logical reason for this: space utilization—the art of getting maximum usable interior space from minimum external dimensions. The Toronado exhibits some improvement in this respect—its flat floors and deep trunk give it superior interior and luggage space to the Buick Riviera, which shares its GM "E" body. However, it's still an extremely large car for its usable space, most of the wasted volume being concentrated ahead of the front wheels.

If the Olds planners didn't worry much about parking and maneuvering in traffic, they did set for themselves some very high standards of stability and roadability. This is among the best-handling big cars we've ever driven. It can be driven through winding mountain roads almost as if it were a sports car, with little lean or protest from the tires—once the front tire pressure is increased to 30 psi. Understeer seems to be quite mild with power on—there's little of the objectionable nose-plowing encountered on comparable cars. If one enters a turn too fast and finds it necessary to back off, the result is a very forgiving characteristic: understeer decreases (it's at a maximum with full power applied), and the car simply gets back on the course you had intended to follow in the first place; without tractive forces at the rear wheels there is no tendency for the rear to break loose. Fine. The only way to really trip this car up in a turn is to *brake,* in addition to backing off. This will bring the tail around, just as it does with any car we can think of. All of these comments apply to driving on wet roads as well as dry; the specially designed TFD tires will actually *squeal* in the rain, indicating very good grip on their part. Anyway, it's our feeling that Olds

engineers have outdone themselves on this aspect of the car, as it not only summarily puts down the traditional objections to fwd on a high-powered car but turns out a better-handling car than its rwd contemporaries.

Behind the wheel the driver has the distinct feeling of piloting some kind of land yacht, and until he gets thoroughly used to the length and width of that snout, he may feel uneasy about driving down city streets. But he does have a very convenient and readable instrument panel in front of him. All instruments and switches are grouped on one panel which stands free of the main dash panel, and everything—including the radio controls—is within easy reach. The drum-type speedometer is one of the most legible we've seen.

The steering wheel is too close to the driver, regardless of the seat position, but there is a tilt-and-telescope wheel available to alleviate this condition. We'd highly recommend that option. Vision forward and to the sides is excellent (except for that long, high hood) and the lack of vent windows provides quite a panoramic effect. Unfortunately vision to the rear is poor—like many current cars in which vision considerations have been outvoted by the stylists. This is so bad as to be hazardous in the Toronado and extreme caution is called for when backing up, or when changing lanes.

In the rear seat there is adequate room for three (as in the front), the flat floor leaving room for good padding in the center. There is a strange, private feeling for passengers in the rear—almost complete isolation from those in the front. The low seats there, the blind rear quarters and the height of the front seatbacks with their headrests all contribute to this feeling. Some will like it and some won't.

Most of our staff had mixed feelings about the Toronado's styling. Whatever the overall esthetic judgment may be, it can't be denied that the styling was carefully planned for effect, from the huge front overhang to the highly emphasized wheels to the sleek tail. It contains at least three elements that foreshadow future U.S. cars; one is the flip-up headlights, another the "chopped" rear end—both of which are destined to become full-blown fads in this country within a year or two—the other is a total elimination of the tradi-

OLDSMOBILE TORONADO
AT A GLANCE...

Price as tested	$5858
Engine	V8, ohv, 6965 cc
Curb weight, lb	4655
Top speed, mph	124
Acceleration, 0–60 mph, sec	9.9
50–70 mph (3–2 gear)	6.0
Average fuel consumption, mpg	11

tional belt line separating the top mass from the lower body mass. The roof flows right into the body and even gives the appearance of being cantilevered from the rear end. If the thin windshield posts could be made transparent, the effect would be complete. This merging of upper and lower body is not a fad, though—simply a logical evolution of body shapes, just as the gradual elimination of fenders was. Handling of the details of surfaces and application of chrome trim to this body have been handled masterfully, and we particularly like the wheels, whose visual interest have been derived directly from a functional requirement—to cool the brakes.

Speaking of brakes—well, maybe we shouldn't. They are a major failing of the car. Total swept area is only 328 sq in., the same as that of pre-disc Corvettes, though there is more lining area. On our brake tests, we managed only a 20 ft/sec/sec deceleration rate on the panic stop, and control over the direction of the car's travel during the stop was a hectic proposition. The problem here seems to be twofold: not enough brake power at the front, where rests probably more than 70% of the car's high weight during such a stop; and too much brake power at the rear. To be exact, the front brakes do 62% of the work, which seems too small a proportion. We can offer some suggestions here: it would seem that discs are a must for the front, for one thing. Non-servo drums at the rear would be a step in the right direction, as they are less prone to lockup (which the present rear ones do very easily). And this car seems to be one of the leading candidates for a pressure limiting device for the rear brakes; most of the world's front-drive cars already have them.

We don't know just what the eventual solution will be. Inboard discs would be relatively easy to fit into the package, but with the extremely high torque loads imposed on the axles and U-joints, Oldsmobile engineers were probably reluctant to add braking torque to the loads on these components. On the other hand, it's a tough proposition to fit discs outboard because of conflicts with steering mechanism. This has been done on some fwd cars, however.

In typical American fashion, the Toronado is at its best on the highway—the ride is gentle, the noise level low. There are no odd noises at all from the power package. As mentioned in our technical analysis, this car has very high spring rates, and these surely play a large part in the excellent handling. They also give a ride that's on the firm side while never being jarring. Much of the refined feeling is lost, though, because of a lack of rigidity in the body structure. There is considerable twisting and creaking on any but the smoothest roads, and our test car was full of rattles that further detracted from an otherwise quiet car. We realize that it's difficult to make a car this large rigid without unit construction or a highly

rigid frame—and we can appreciate Detroit's aversion to the ride harshness and road rumble that can result from unit bodies. But it seems to us that the very gains realized from these separate bodies and flexible perimeter frames—isolation from harshness and rumble—are more than lost to the rattles and squeaks that result.

In line with the car's superhighway capability, it has directional stability of an order we've never seen surpassed. This is good, for with nose-heavy rear-drive cars the requirements of straight-line directional stability, moderate understeer and good traction under slick road conditions have been antagonistic: increased weight up front gives the desired directional stability but increases understeer beyond desirable limits and defeats traction. With fwd the three requirements seem to have been made compatible with each other. With its wonderful immunity to side winds, the Toronado should be soothing to drive when others are being buffeted about.

In summary, then, front-wheel drive seems to work extremely well for the large American car. Its full benefits won't be realized until the driver has met with some adverse conditions—mud, snow, side winds, winding roads. With this car Oldsmobile can gain production experience with the layout, gradually reduce the cost penalty involved and get it into their less extravagant family cars. Properly used, fwd could revolutionize the American car by allowing it to be much smaller in overall size and more stable, without sacrificing space or comfort. Maybe this is what they have in mind for the future. As it stands, the Toronado is like most American cars, with all the virtues and vices that entails—but with superior handling under adverse road conditions, and slightly better space utilization.

There may be a few readers who aren't thoroughly familiar with the Toronado; for their benefit we shall review the car's layout briefly. The first American car to have front-wheel drive since the Cord 812 was discontinued in 1937, it is also the world's largest and most powerful car to feature fwd.

In overall length and width the Toronado is about the same as the American "full-size, low-price" sedans—typified by the large Fords, Chevrolets and Plymouths. At 4655 lb, it's quite a bit heavier than any of these cars and more on a par with the "luxury personal" cars with which it is intended to compete in the marketplace, Thunderbird and Riviera.

Power is furnished by a full 7-liter ohv V-8, developing an advertised 385 bhp and positioned longitudinally in the car. From this giant of an engine torque is then taken through a torque converter with a two-position stator to the drive sprocket of the transfer case. A two-inch-wide rubber-cushioned chain takes the drive down and to the left, to the input end of the three-speed automatic planetary gearbox of the same type as used in other GM cars. From here a planetary differential delivers the torque to the half-shafts that drive the front wheels. Front suspension is by A-arms and torsion bars, rear by a beam axle and single-leaf springs.

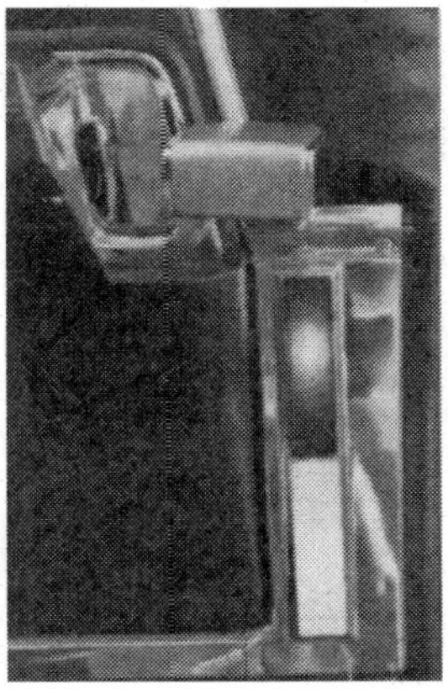

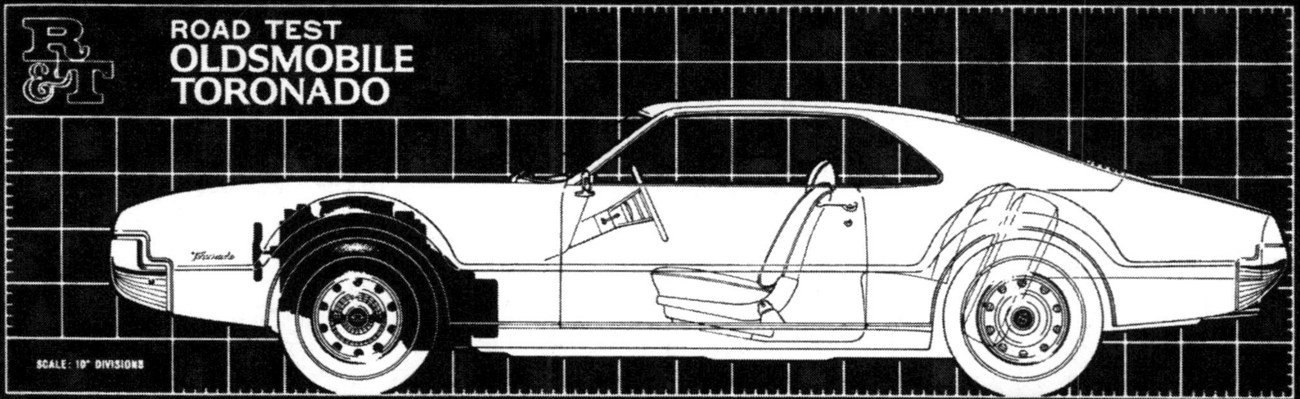

PRICE

Basic list $4812
As tested $5858

ENGINE

No. cyl. & type V-8, ohv
Bore x stroke, mm 105 x 101
In. 4.11 x 3.98
Displacement, cc/cu in 6965/425
Compression ratio 10.5:1
Bhp @ rpm 385 @ 4800
Equivalent mph 124
Torque @ rpm, lb-ft 475 @ 3200
Equivalent mph 83
Carburetors, no. & make 1 Rochester
No. barrels, dia 2-1.375 2-2.250
Type fuel required premium

DRIVE TRAIN

Transmission: automatic, torque converter + 3 speeds
Gear ratios: 3rd (1.00) 3.21:1
2nd (1.48) 4.76:1
1st (2.48) 7.96:1
1st (2.2 x 2.48) 17.6:1
Differential type planetary
Ratio 3.21

CHASSIS & SUSPENSION

Frame type: boxed perimeter terminating at rear spring eye.
Brake type drum
Swept area, sq in 328
Tire size 8.85-15
Make & model Firestone TFD
Steering type power
Turns, lock-to-lock 3.4
Turning circle, ft 43.3
Front suspension: unequal A-arms, torsion bars, tube shocks, anti-roll bar.
Rear suspension: beam axle, single-leaf springs, vertical and horizontal tube shocks.

ACCOMMODATION

Normal capacity, persons 4
Occasional capacity 6
Seat width, front, in 55.0
Rear 54.0
Head room, front/rear 37.0/35.0
Seat back adjustment, deg 0
Entrance height, in 51.0
Step-over height 13.0
Door width 46.2
Driver comfort rating:
Driver 69 in. tall 85
Driver 72 in. tall 80
Driver 75 in. tall 80
(85–100, good; 70–85, fair; under 70, poor)

GENERAL

Curb weight, lb 4655
Test weight 5035
Weight distribution (with driver), front/rear % 62/38
Wheelbase, in 119.0
Track, front/rear 63.5/63.0
Overall length, in 211.0
width 78.5
height 52.8
Frontal area, sq ft 23.0
Ground clearance, in 4.7
Overhang, front/rear 48.0/42.0
Departure angle, deg 13.4
Usable trunk space, cu ft 14.5
Fuel tank capacity, gal 24.0

INSTRUMENTATION

Instruments: 130-mph speedometer, fuel level, water temp, ammeter, clock.
Warning lights: oil pressure, high beam, directionals, parking brake.

MISCELLANEOUS

Body styles available: 2-door hardtop (as tested).

EXTRA COST OPTIONS

No power or drive-train options. Convenience items too numerous to list.

CALCULATED DATA

Lb/hp (test wt) 13.1
Mph/1000 rpm (3rd gear) 25.8
Engine revs/mi 2320
Piston travel, ft/mi 1540
Rpm @ 2500 ft/min 3780
Equivalent mph 97.5
Cu ft/ton mi 113.0
R & T Wear Index 35.7

MAINTENANCE

Crankcase capacity, qt 6.0
Change interval, mi 6000
Oil filter type full flow
Chassis lube interval, mi 12,000

FUEL CONSUMPTION

Normal driving, mpg 10–13
Cruising range, mi 240–312

ROAD TEST RESULTS

ACCELERATION

Time to speed, sec
0–30 mph 3.2
0–40 mph 4.8
0–50 mph 7.1
0–60 mph 9.9
0–70 mph 13.1
0–80 mph 16.6
0–100 mph 26.0
50–70 mph (3–2 gear) 6.0

Time to distance, sec
0–100 ft 4.0
0–500 ft 10.0
¼ mile 17.8
Speed at end, mph 83
Passing exposure, sec 6.6

SPEEDS IN GEARS

High gear (4800), mph 124
2nd (4700) 80
1st (4700) 48

BRAKES

Panic stop from 80 mph
Deceleration rate, ft/sec/sec 20
Control fair
Parking: hold 30% grade no
Overall brake rating poor

SPEEDOMETER ERROR

30 mph indicated actual 29.8
40 mph 39.3
60 mph 57.5
80 mph 75.6
100 mph 92.7
Odometer correction 0.952

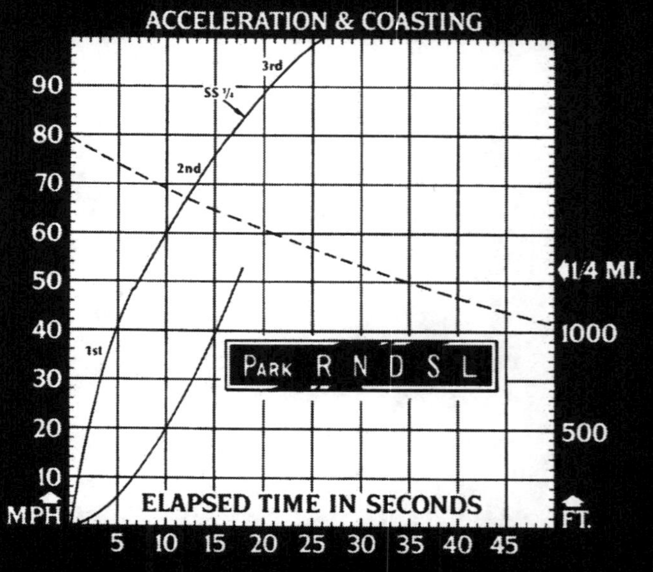

Oldsmobile Toronado 6,965 c.c.

Autocar Road Test
NUMBER 2061

MANUFACTURER
Oldsmobile Division, General Motors Corporation, Lansing, Michigan, U.S.A.

BRITISH CONCESSIONAIRES
Lendrum and Hartman Ltd., Flood Street, London, S.W.13.

PRICES

Basic	£3,365 0s 0d
Purchase Tax	£702 12s 1d
Total (in G.B.)	£4,067 12s 1d

(Includes radio, turning lamps, electric seat adjusters and window lifts, "tilt and travel" steering-wheel, safety harness, etc.).

EXTRAS (INC. P.T.)

Air-conditioning	£290
Reclining passenger seat and head-rests	£58

PERFORMANCE SUMMARY

Mean maximum speed	126 m.p.h.
Standing start ¼-mile	16·9 sec
0-60 m.p.h.	8·7 sec
30-70 m.p.h. in intermediate	9·7 sec
Overall fuel consumption	11·8 m.p.g.
Miles per tankful	240

AT A GLANCE: America's only front-drive car shows superior traction and roadholding properties, combines powerful acceleration and 126 m.p.h. limit with extraordinary quietness and mechanical refinement. Excellent high-geared power steering, but needs better brakes. Softly damped but roll-free suspension. 4-6-seater coupé body with all mod. cons., a real eyecatcher.

OF all the exciting new things to be seen at the international motor shows last autumn, none attracted closer scrutiny among the technical cognoscenti—nor perhaps drew more scepticism—than the 7-litre Oldsmobile Toronado, largest and most powerful production car ever to be driven by its front wheels. Although the great General Motors combine had brought off revolutionary coups before (the Chevrolet Corvair, for example, with its flat-six air-cooled engine in the tail), it has also muffed one or two. Even allowing for the "boardroom factor" in the 385 advertised b.h.p., how could this new car possibly compete in New York's block-to-block races between traffic signals? If one were to shut off power suddenly midway through a corner, would an understeer-oversteer transition make it almost unmanageable for ordinary mortals?

Appreciating the exceptional interest in the Toronado, G.M.'s very obliging representative in London has lost no time in having an example sent over here for the Press to try. After covering nearly 2,000 miles in it under varying conditions, both in this country and on the Continent, we judge its general road behaviour clearly superior to that of any equivalent rear-wheel drive product, and predict that other transatlantic manufacturers must inevitably follow suit.

Probably it was to make sure that people would buy the Toronado despite any doubts or prejudices about f.w.d. that G.M. styled it so boldly, and indeed took no real advantage of this layout's benefits in "space utilization." The enormous frontal overhang, for instance, is strictly non-functional since there are no mechanical components—other than the fan—forward of the wheels. With the widest tracks of almost any current production car (5ft 3·5in. front, 5ft 3in. rear), overall width of 6ft 6·5in. and 17ft 7in. from bumper to bumper, it is truly massive for a 2-door coupé, and the styling is both bold and original.

What other car would entice a policeman on point duty to leave his post at a busy moment, stroll over and say: "Are you in the right country, sir—or even in the right

Autocar Road Test 2061

MAKE: Oldsmobile

TYPE: Toronado

WEIGHT

Kerb weight (with oil, water and half-full fuel tank): 40·8cwt (4,570lb-2,073kg)
Front-rear distribution, per cent .. F, 61·1; R, 38·9
Laden as tested .. 43·8cwt (4,906lb-2·225kg)

TURNING CIRCLES

Between kerbs .. L, 42ft 9in.; R, 44ft 0in.
Between walls .. L, 45ft 10in.; R, 47ft 0in.
Steering wheel turns to lock .. .. 3·4

PERFORMANCE DATA

Top gear m.p.h. per 1,000 r.p.m. .. .. 25·8
Mean piston speed at max. power .. 3,184ft/min
Engine revs at mean max. speed .. 4,880 r.p.m.
B.h.p. per ton laden (gross) 176

OIL CONSUMPTION

Miles per pint (SAE 20W) 300

FUEL CONSUMPTION

At constant speeds

30 m.p.h.	19·5 m.p.g.	70 m.p.h.	16·8 m.p.g.
40 ,,	20·2 ,,	80 ,,	15·6 ,,
50 ,,	20·0 ,,	90 ,,	13·6 ,,
60 ,,	18·5 ,,	100 ,,	11·4 ,,

Overall m.p.g. .. **11·8 (24·0 litres/100km)**
Normal range m.p.g. 10-15 (28·3-18·8 litres/100km)
Test distance .. .. 1,850 miles
Estimated (DIN) m.p.g. 15·3 (18·5 litres/100km)
Grade .. .. Super premium (100·3-101·8RM)

TEST CONDITIONS

Weather .. Occasional light drizzle with 0-5 m.p.h. wind
Temperature .. .. 9 deg. C. (48 deg. F.)
Barometer .. .. .. 29·8in. Hg.
Mostly damp concrete and tarmac surfaces.

Speed range, gear ratios and time in seconds

m.p.h.	Top (3·21—7·06)	Inter (4·75—10·46)	Low (7·96—17·49)
10— 30	—	—	2·7
20— 40	6·0	4·1	2·9
30— 50	6·2	4·2	3·3
40— 60	6·6	4·9	—
50— 70	7·3	5·5	—
60— 80	9·1	6·2	—
70— 90	9·8	—	—
80—100	10·6	—	—
90—110	11·6	—	—

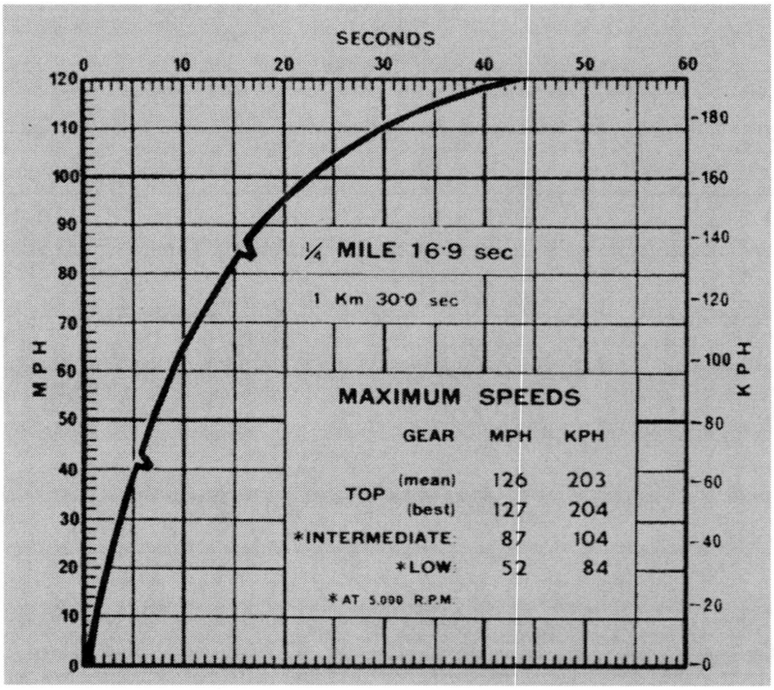

TIME IN SECONDS	3·8	5·1	6·7	8·7	11·3	14·0	17·7	22·5	29·3	
TRUE SPEED MPH	30	40	50	60	70	80	90	100	110	120
INDICATED MPH	31	42	52	62	72	83	93	103	113	

BRAKES (from 30 m.p.h. in neutral)	Pedal load	Retardation	Equiv. distance
	25lb	0·22g	137ft
	50lb	0·45g	67ft
	75lb	0·87g	34·6ft
Parking brake		0·20g	150ft

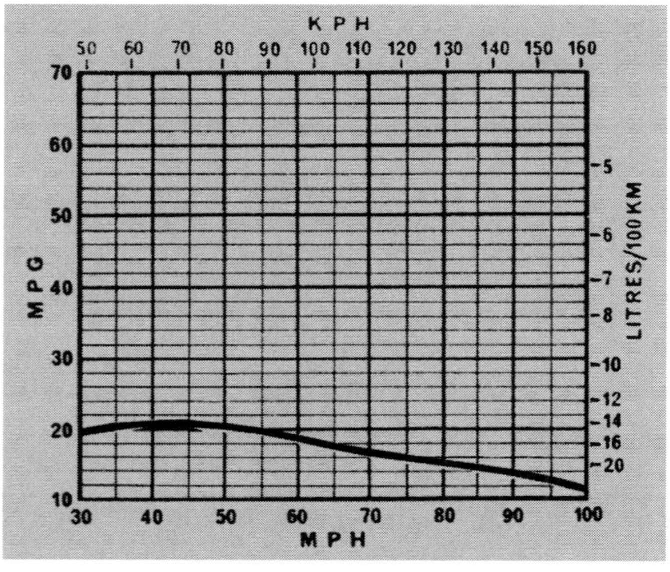

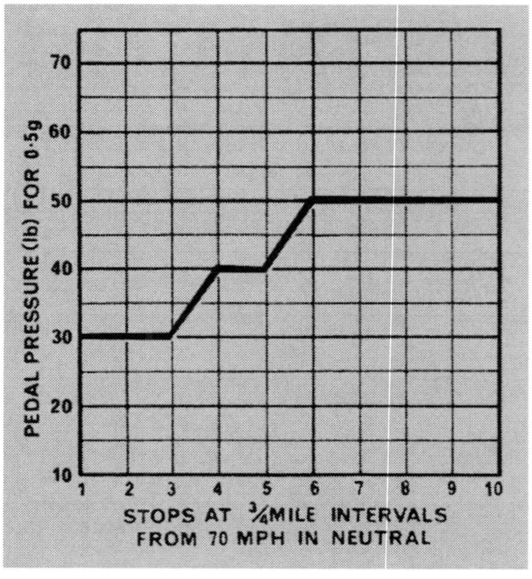

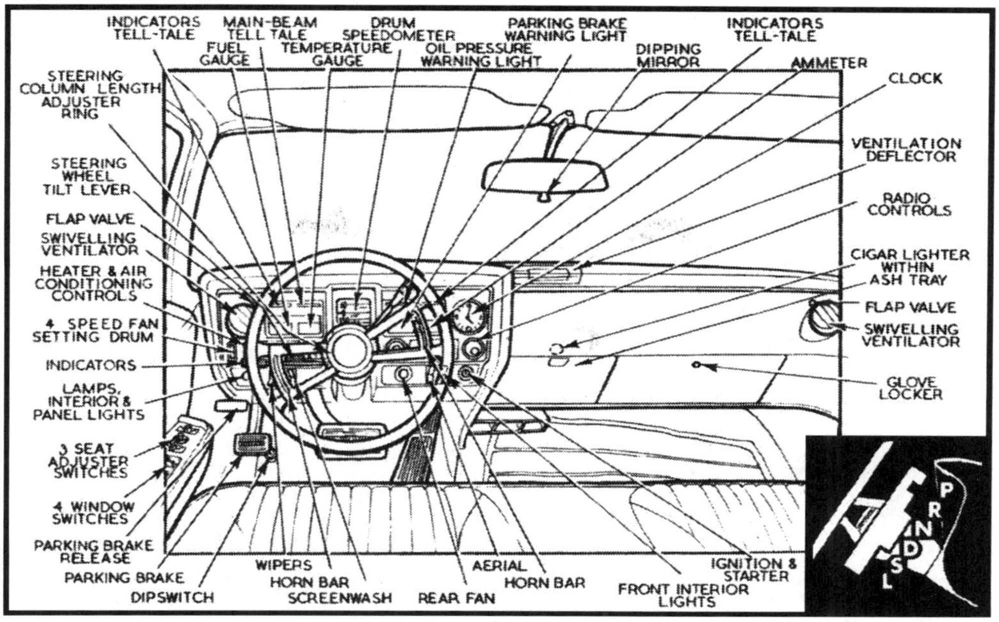

world?", or the driver of a lorry halted at a traffic signal to leap from his cab to come and ask about it?

Before starting up the Oldsmobile and getting going, so to speak, perhaps we should run through its main features for those readers unfamiliar with the design. It has a separate, box-section perimeter frame extending back only as far as the forward eyes of the single-leaf rear springs, a subframe welded to the underside of the body supporting it from this point back. The front wheel assemblies are carried on wishbones and sprung by longitudinal torsion bars the back axle being a rigid beam.

Offset to the right of the car's centre line, the big vee-8 engine has a two-stage torque converter on the tail-end of the crankshaft, and behind this a toothed chain-and-sprocket drive to a 3-speed automatic gearbox lying to the left of, and parallel with, the crankcase. From this the drive is taken forward to a hypoid bevel drive with planetary differential. Drum brakes, with self-wrapping shoes in radially finned light alloy drums, are all outboard.

From the driving seat one could be somewhat overawed at first by the length and breadth of the great prow, which has no fall-away to give one a sight of the road close-to; this is a particular disadvantage in fog, of course. The multiplicity of controls is likewise rather forbidding until one learns their purposes and whereabouts, but in fact their arrangement is excellent, as is the use of different types of switch for the various functions, by which they are memorized very quickly.

Having studied the cockpit layout and drill, the driver adjusts his seating position for height and tilt as well as reach, using three switches set in the door. On the test car the relationship between cushion and backrest was fixed on the driving seat, but the passenger's had a lever-release adjustment for the backrest. Then the rake of the steering-wheel can be set to any of six angles, from " Mini " to near-vertical, by simply lifting a short lever on the column. The hub serves as a locking ring for a spring-loaded telescopic adjustment of the column, with a 3in. range. This is extremely easy to operate, needing very little effort to unlock or lock securely.

Whatever the weather, one simply presses the accelerator to the floor once, releases it and turns the ignition key for the first start of a day. The engine fires up right away, after which, apart from the possibility of a rick stall if one did not clear it with a burst of revs before setting off, there were no flat-spots or other carburation shortcomings. For hot starts the accelerator is pushed about halfway down before turning the key. There are no vibration periods right up to peak revs, and one could not imagine the engine being quieter or smoother with double the number of cylinders.

Although we are now accustomed to American engines, this one—combined with a transmission virtually silent at any speed and in any gear—seemed to us the quietest yet. In fact, its almost uncanny freedom from mechanical commotions or wind roar around the body, right up to the maximum of 126 m.p.h., is perhaps the Toronado's outstanding virtue.

It seems improbable that the average American motorist who rarely if ever extends his car to the full would detect any difference between this and the conventional type in normal conditions. No pull or other influence comes back through the powered steering to the driver's hands, even on full lock; in fact, it might be better if there were a little less assistance and more "feel." But where a wet road might find him embarrassed with spinning wheels and sliding tail in a rear-driven car, the Oldsmobile's superior traction would get him off the mark much more smartly and with much more directional control.

From this level one cannot appreciate the engine's offset, but the top of the chaincase behind the torque converter is just in view. Belt-driven alternator and steering hydraulic pump are mounted high, and all engine fumes are drawn into the induction

Oldsmobile Toronado ...

When we pointed the Toronado nose upwards on the 1-in-3 test hill, with the concrete damp, we expected to be left standing. Yet, no doubt assisted by the gentle power delivery of the hydraulic torque converter, it moved off at once without wheelspin. A pointer to the steepness of 1-in-3 is that, with the nose down and parking brake applied, the car slid forward with its rear wheels locked. This brake was not man enough to hold the car the other way round, but could just cope with 1-in-4.

All our standing-start acceleration figures had to be taken on a damp surface, although certainly one with a good frictional coefficient. With full power applied from the word "go" there was astonishingly little wheelspin, and the consistency of the resulting figures was remarkable. In eight successive runs the elapsed time for covering a kilometre from rest varied by only 0·3 sec. On *Autocar's* electric speedometer the quarter-mile post was passed at 88 m.p.h., the half-mile at 106 and kilometre at 112.

It is interesting to compare the Toronado's performance with that of the Buick Riviera tested last year. The engine capacities are almost identical, but the Riviera had a larger bore and shorter stroke, produced peak power at lower revs and yet was lower-geared. Running in perfect weather conditions, the Riviera was quicker off the mark, reaching 40 m.p.h. in the time it took the Toronado to reach 30, after which none of its figures up to 110 m.p.h. are more than 1½ sec slower. But the Riviera's single-gear figures were appreciably better at lower speeds, where its engine torque was stronger. The Toronado, on the other hand, is a few m.p.h. quicker and used less fuel. All the same, one does not set off for a week-end's motoring without a few fivers for super premium.

On fast straight roads the Toronado is stable as an arrow, having over 60 per cent of its 2¼ tons on the front wheels. Driven normally, it runs very easily and accurately round corners—once the driver has learned not to overdo the "one-finger" steering—and almost without roll. It is not until one is deliberately looking for tricks, by cutting the throttles suddenly in a bend, that the front-drive characteristics become quite prominent. With the throttles open the front tends to run wild; when they are closed the tyre slip angle is reduced and the nose turns in, as with smaller f.w.d. cars. But the transition is neither violent nor unstable, and one can use it to advantage. The ultimate cornering power is extremely high, and on a closed circuit we were leaving black lines on the road beneath the inner front wheel, so much

The doors are necessarily rather wide and heavy; seat and window switches are grouped in the driver's. Front-drive allows flat floors, and lap-strap safety belts are standard front and rear. The steering-wheel is adjustable for rake and reach

An armrest dividing the front seat folds up to complete the backrest when three are carried. The passenger's reclining backrest is infinitely adjustable between its limits. Side windows lower flush without a dividing pillar

Although the boot is wide and deep, it needs an acrobat to remove or replace the spare wheel. Fluted strips beside the number plate are dazzling brake lamps; above these are small amber flashers to comply with British law. Reversing lamps are set in the deep, rugged bumper

power could one apply without ploughing straight on.

Although our test period was almost free from ice, and completely without snow, obviously the Toronado would be much more stable and have better traction in such conditions than its conventional compatriots. We did manage once to sink one front wheel very deep in Belgian mud while turning, and were grateful for its self-extrication. The Toronado's tyres are specials by Firestone; their behaviour is extremely creditable under all conditions, and from inside the car with the windows shut one scarcely hears them. With a window open while passing between buildings, though, there is a pretty impressive roar.

As with most American cars, the Achilles heel of this one lies in its braking system, which for British and European road conditions is quite inadequate. One can get away with using them little and often, since their cooling is apparently sufficient for this, but one full-energy stop from 90-100 m.p.h. has them "cooked" beyond capacity long before the car has come to rest. However, their recovery is very rapid. At low speeds they are rather fierce, and need to be more progressive. Even in dry weather the need for an abrupt stop can be a bit unnerving, as the self-wrapping shoe action is liable to lock the back wheels and slew the tail. Since there is no pressure limiting valve between front and rear systems, the braking stability is at its best with the petrol tank full and a good load on board.

Fortunately this big and heavy car's handling characteristics are such that one can move very quickly and safely from A to B without depending much on the brakes, and the transmission allows one to change down for extra engine braking from any speed. The governor being inoperative with S (for Super, alias intermediate) selected, one must guard against doing this above 85-90 m.p.h. Low gear also has an ungoverned manual hold. When performance testing, incidentally, best results were achieved with the automatic change-point from low to intermediate, but holding the latter gear (with the selector at S) up to about 85 m.p.h., or 5,000 r.p.m.

One rides very comfortably on springs that are a bit firmer than the American norm, but most Europeans would prefer stiffer damping of the front suspension. As it is, the nose heaves slowly up and down in nautical fashion much of the time, sometimes with a slight corkscrew motion, and after a big spring deflection the movement takes several seconds to die out. Back-seat passengers ride as comfortably as those in front, but sit rather low and thus have a reduced view of the landscape rushing past them. The optional (and detachable) front headrests are quite a blind-spot for them. It's rather like sitting behind the pilot in a modern light cabin plane, the quietness and sense of detachment as well as the rather out-of-this-world interior decor accentuating this impression.

Over the washboard and *pavé* at the MIRA proving grounds the Oldsmobile performed competently if without special distinction, body shake being less than expected since on ordinary roads our car was plagued with some creaking behind the facia. We were assured that this could easily be silenced.

Reverting a moment to the steering, one must criticize the rather clumsy turning circles—well over 40ft between kerbs—but give credit for the combination of a small (15½in.) wheel and relatively high gearing.

As the vacuum-operated lifters for the dolly's eyelid headlamps take about 6-8sec to do their job, one has to anticipate the need for them at times, especially for motorway signalling. There is no flasher control, and the light intensity is none too strong on either high or low beams. As we approached maximum speed in daylight, incidentally, one eyelid popped up for a peep. Greatly appreciated were the turning lights in the front wings synchronized with the direction signals; they can be especially useful in dense fog. Red and white (safety and courtesy) lamps are set in the inner panels of each door. It is good to know that, in addition to a dozen fuses for the electrics, there are also circuit breakers

TOTAL PRICE £4,068
£3,929
£4,424
£2,355
£2,340

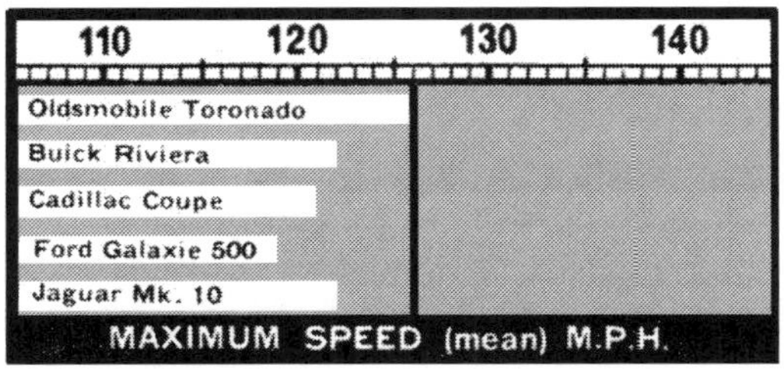

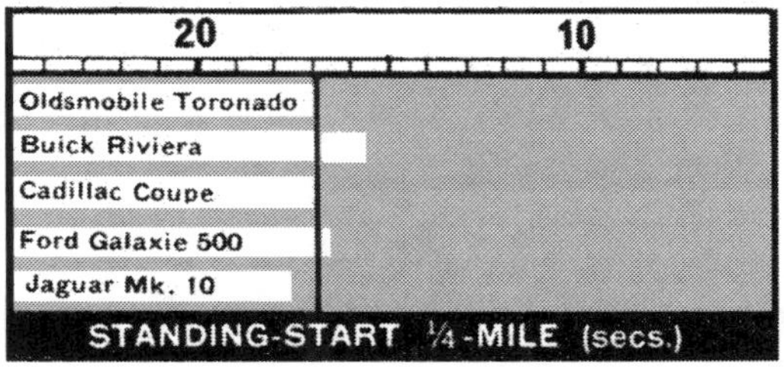

HOW THE OLDSMOBILE TORONADO COMPARES:

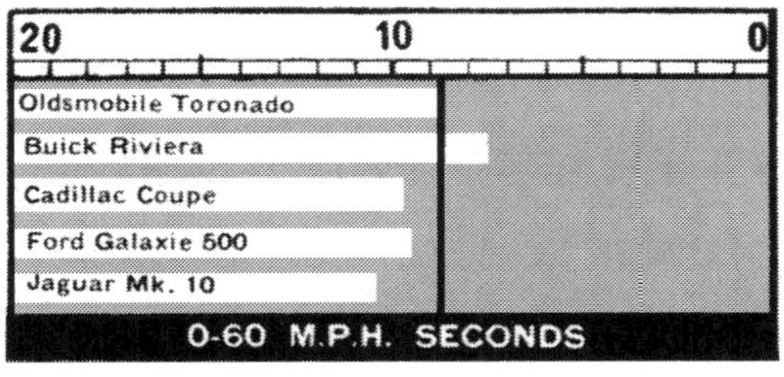

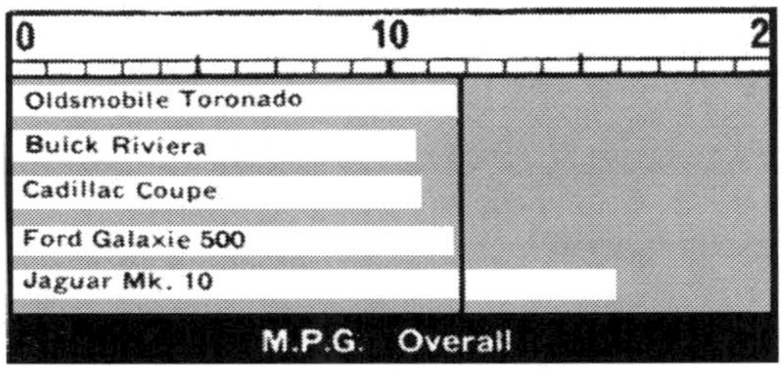

for the most important ones.

An innovation for an American car is the drum-shaped speedometer with horizontal spindle. It was commendably steady and near-accurate, but not so easy to interpret at a glance as a needle on a dial. Among the many luxuries and detail refinements are screenwashers squirting intermittently, the wipers automatically starting in conjunction with them. The two-speed wipers were very effective up to about 85 m.p.h., after which they began to paw the air.

There is a sad lack of stowage for maps and oddments—simply a small locker in the facia, and not even a shelf behind the back seat. This seems especially strange, coming from a country of born travellers with immense distances to cover between trading and holiday centres. The heating and ventilation system on our car included refrigeration (which we did not use). It had an extremely powerful 4-speed blower for the front and two-speed demister for the back window. Apart from a temperature setting lever, there were six push-buttons for the various functions, but one could not admit cool air above while keeping one's feet warm. However, a tropical greenhouse atmosphere could be reached very soon after a cold start. In addition to all the gadgets installed, the list of other intriguing options would be too long to quote. Scheduled servicing requirements are extremely few, in keeping with the current trend. All in all, this Oldsmobile is a sort of dream car that lives up to its exotic appearance in most respects, and that in one blow destroys any illusions about front-drive being unsuitable for very large and powerful vehicles. Even with a right-hand conversion it (since the manufacturers do not make it that way) it would still be rather bulky for our roads, but nevertheless a very desirable property. ■

SPECIFICATION : OLDSMOBILE TORONADO FRONT ENGINE, FRONT WHEEL DRIVE

ENGINE

Cylinders	8 in 90 deg. vee
Cooling system	Water; pump, thermostat and fan with viscous coupling
Bore	104·8mm (4·12in.)
Stroke	101·0mm (3·98in.)
Displacement	6,965 c.c. (425 cu. in.)
Valve gear	Overhead, pushrods, hydraulic tappets and rockers
Compression ratio	10·5-to-1
Carburettor	Rochester 4-choke progressive
Fuel pump	Mechanical
Oil filter	Full-flow renewable element
Max. power	385 b.h.p. (gross) at 4,800 r.p.m.
Max. torque	475 lb. ft. (gross) at 3,200 r.p.m.

TRANSMISSION

Gearbox	Turbo Hydra-Matic 3-speed with torque converter
Gear ratios	Top 1·0-2·2; Inter. 1·48-3·26; Low 2·48-5·45; Reverse 2·08-4·58.
Final drive	Hypoid bevel 3·21 to 1

CHASSIS AND BODY

Construction	Separate perimeter frame, steel body

SUSPENSION

Front	Independent, torsion bars, wishbones, anti-roll bar, telescopic dampers
Rear	Dead beam axle, single leaf springs, vertical and horizontal telescopic dampers

STEERING

Type	Power-assisted Saginaw, recirculating ball. Turns lock to lock 3·4. Wheel dia. 15·5in.

BRAKES

Make and type	Self-adjusting, self-wrapping drums
Servo	Bendix vacuum
Dimensions	F, 11in. dia.; 2·75in. wide shoes R, 11in. dia.; 2·0in. wide shoes
Swept area	F, 190·0 sq. in.; R, 138·2 sq. in. Total 328 sq. in. (150 sq. in. per ton laden)

WHEELS

Type	Ventilated cast aluminium, 5 studs; 6in. wide rim
Tyres	Firestone T-FD tubeless, size 8·85-15in.

EQUIPMENT

Battery	12-volt 73-amp hr. negative earth
Alternator	Delco Remy 42 amp.
Headlamps	Main beams—4 × 37·5-watt. Low beams—2 × 50-watt
Reversing lamp	2 standard
Electric fuses	12, plus 3 circuit breakers
Screen wipers	2-speed self-parking
Screen washer	Standard, electric
Interior heater	Standard, 4-speed fan
Safety belts	Standard, lap straps front and rear
Interior trim	P.v.c. seats, p.v.c. headlining
Floor covering	Carpet, with rubber mats in front
Starting handle	No provision
Jack	Vertical pillar with ratchet lever
Jacking points	2 under each bumper
Other bodies	None

MAINTENANCE

Fuel tank	20 Imp. gallons (no reserve) (91 litres)
Cooling system	28·3 pints (including heater) (16 litres)
Engine sump (inc. filter)	10 pints (5·7 litres) SAE 10W-30 or 20W. Change oil every 2 months (or 6,000 miles if covered in shorter period). Change filter element every 6 months or 6,000 miles
Torque converter, gearbox and final drive	20 pints GM-1050015. No changes
Grease	Steering and suspension ball-joints every 12,000-36,000 miles
Tyre pressures	F, 24; R, 24 p.s.i. (normal driving); F, 28; R, 28 p.s.i. (fast driving); F, 26; R, 26 p.s.i. (full load); F, 30; R, 30 p.s.i. (full load and fast driving)

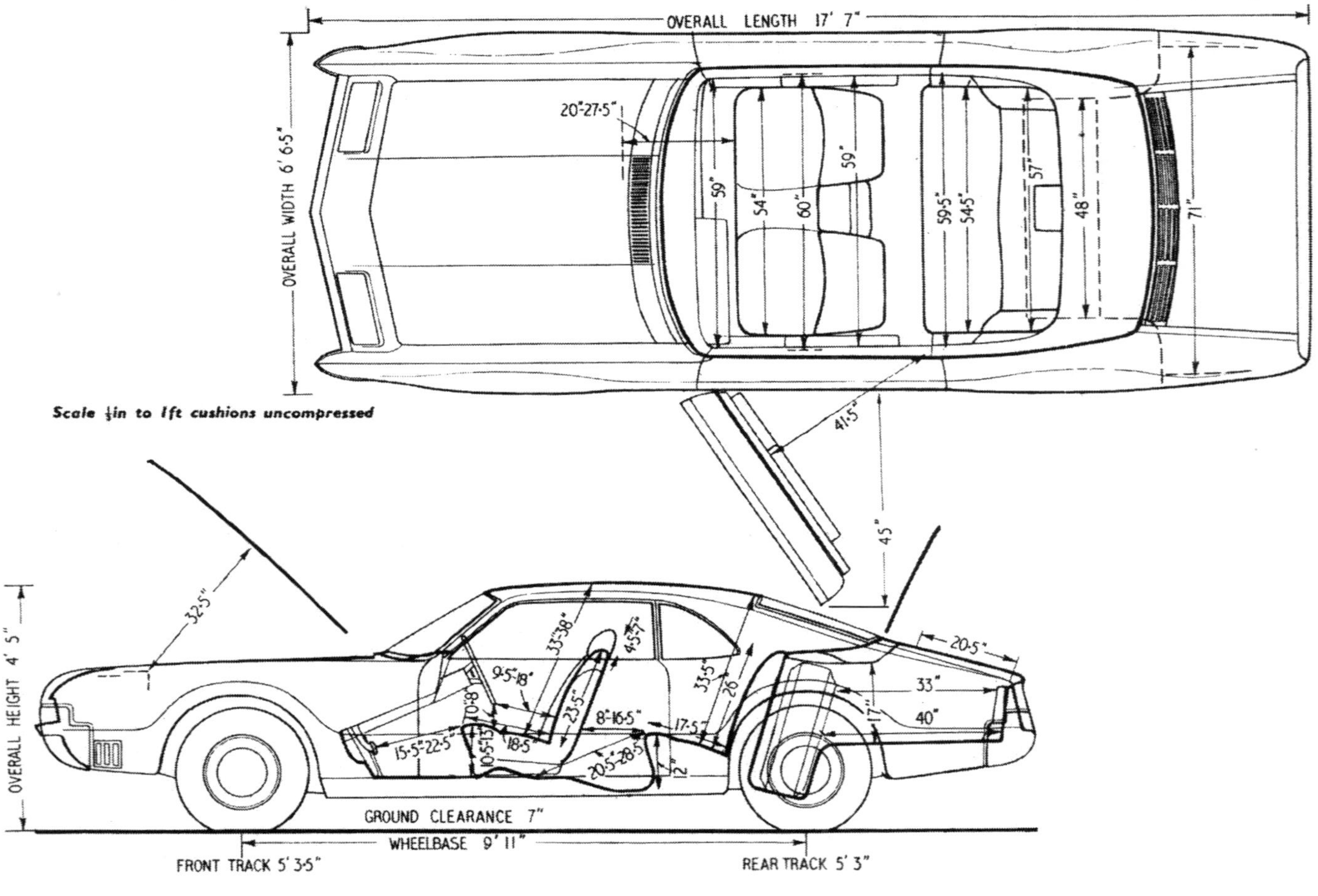

Scale ¼in to 1ft cushions uncompressed

TORONADO AT THE PEAK

Oldsmobile builds the Toronado's performance image with a rapid run up Pikes Peak

BY ROGER HUNTINGTON

There's an old wives' tale to the effect that a front-wheel-drive car can't climb hills. Not on loose gravel anyway. The old wives in question say that the rearward weight transfer—resulting both from the car's nose-up attitude and from the reaction to acceleration—un-loads the driven front wheels to such an extent that they can't get any traction, and the tires just claw helplessly at the road. When Oldsmobile brought out their front-wheel-drive Toronado, the fellows in Lansing, Michigan, were hoping it wouldn't be true of this car. Realizing that this design would be fair game for the critics, Oldsmobile decided to prove the Toronado's practicality for everyday driving (and score a few points for the ol' performance image too) by having a shot at that infamous stretch of gravel, sand and snow that snakes up Pikes Peak.

The idea for the hill-climbing run was Ted Louckes', the assistant head of experimental engineering at Oldsmobile. He was in a position to undertake this project without ruffling any anti-racing feathers on GM's Mahogany Row in Detroit. It wasn't a race anyway—only against the clock. And the stunt seemed like a natural to prove some of the tremendous traction potential of this fwd design.

It's history now that the run was a complete success. Driver Bobby Unser turned the wild 12-mile course in 14:09.9—which was within 17 seconds of Parnelli Jones' all-out stock car record in a Mercury two years ago. Traction was all that was hoped for, but the project taught Olds engineers lessons about the handling and roadability of their new toy in rough mountain driving that would've required months of proving-ground testing to learn.

And don't minimize Unser's part in the game. He was more than just a skilled driver who knew the hill. "We couldn't have tackled this without Bobby," says Louckes. "His knowledge of setting up a car for the hill was as valuable as his driving ability. Our boys hardly knew where to start to modify for traction on gravel—we would've done almost exactly opposite to what Bobby suggested—and it would have required literally weeks of testing to sort things out by trial and error. Also, Bobby's ability as a test driver amazed us. He could run a certain section of the course at times that wouldn't vary more than a couple of tenths—and then tell us exactly what had to be done to the suspension to get absolute minimum time. He drives at ten-tenths all the time on that hill. And there was no question in his mind when the car was right: he said, 'That's it,' and was perfectly confident that there was no way to get that particular car up there any faster. The kid is something else."

Setting up the Toronado suspension for maximum bite on those gravel turns was probably the most interesting part of the project. It's just the opposite to what you would do on dry pavement, because the effective traction coefficient is so much lower on the gravel. It's almost like setting up a low-powered car to get around on ice. The key here is to *increase* body roll! On the Toronado, they ended up using a smaller front anti-roll torsion bar and 25 per cent softer springs and shocks all around. The anti-roll bar they brought along to put on the rear proved to be a hindrance. The suspension ended up absolutely flabby.

Picture the effect of this flabby suspension on a gravel turn. What they were doing, in effect, was to let the work required to roll the sprung mass absorb some of the centrifugal force that would otherwise lift the inside wheels. Keep in mind that this is strictly a *momentary* effect—what engineers call a "transient" condition. In a long high-speed turn, where the body settles down in a fixed attitude, this hefty roll effect would *hurt* the cornering power. But when you're wheeling a heavy car back and forth through tight switchback turns, body roll can have a big effect on instantaneous wheel loads. In a word, when the car is wallowing around, the tires are staying on the ground. The roll inertia of the sprung mass is absorbing centrifugal force, at least for that first second or so as the car enters a turn. And on tight turns like these, the car is almost through the turn before the sprung mass settles down. Think about that a second. If they had set the Toronado up with tight suspension and minimum roll, the inside tires would have been sliding the split-second the driver changed direction.

Oldsmobile engineers didn't know about this. Bobby Unser did. Unser also knows, as do all dirt track drivers, that maximum traction is obtained by transferring as much weight as possible to the outside wheels. In dirt track cars, this is accomplished by building them tall; with a high center of gravity, centrifugal force presses the outside wheel onto the surface. In a car with a low center of gravity, the only way such lateral weight transfer can be accomplished is by having such a loose suspension that the body heels over and leans extra weight on the outside wheels. So the flabby suspension of the Pikes Peak Toronado had a double beneficial effect: good transient response and better weight transfer, but the latter comes into effect after the transient condition is over and the car has settled into a stable attitude, as during a long, sweeping turn.

Climbing a hill is something like acceleration in that the car's weight-to-horsepower ratio is the prime performance parameter (except that you're lifting a mass as well as accelerating it). Thus Olds' major aim on the Pikes Peak Toronado before it left Lansing was to cut the weight as much as possible, and hop up the engine a little—within the USAC rules defining a "stock" engine. The weight-cutting program was amazing—they took out more than 1000 pounds. The body was completely stripped and gutted, even to many of the body bracing members. And a lot of weight was taken out of the frame. They ran a "Stresscoat" test first, applying a thin plastic coating on all frame members, then subjecting the frame to normal running stresses. Points of maximum stress can be easily read by noting the cracks in the plastic coating. So Olds engineers just whacked out the areas of minimum stress. Frame weight was cut over 100 pounds without seriously weakening it. The steel hood, front bumper and rear deck lid were replaced with fiberglass duplicates, and the heavy front and rear windows were replaced by plexiglass. The inner fender panels were "Swiss-cheesed." The instrument panel was replaced by a few rudimentary gauges on an aluminum panel, and the seat was just a light fiberglass bucket.

They managed to get the weight down to

Veteran Pikes Peak competitor Bobby Unser checks his time after the run. He climbed the hill just 17 seconds slower than the stock car record.

This was no stock Toronado. The interior was gutted and liberal amounts of weight were removed, in accordance with USAC stock car regulations.

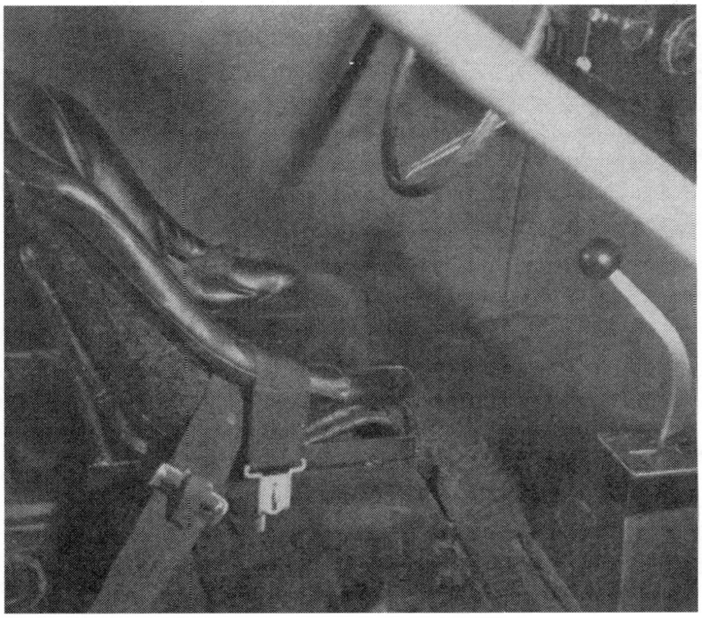

BOBBY UNSER-DRIVER

3400 pounds from 4500, and slightly increased the percentage of weight on the front driving wheels, from 60 per cent to 62-63. All to the good.

The engine was a routine hop-up job. USAC rules say you can do anything inside, as long as you're under 427 cu. in., and carburetion can be only a single four-barrel. Modifications included a hotter experimental hydraulic cam that would permit 5500 rpm in the gears, milling the heads for 12-to-one compression, porting and polishing, increased clearances, careful balancing of the lower end, and a good set of tuned exhaust headers of steel tubing (which save weight as well as increase horsepower). Dyno tests showed a gross peak of about 410 bhp at 5200 rpm on premium pump gas—which is well over 100 bhp more than the standard Toronado engine produces as installed. (Advertised horsepower doesn't mean much outside the dyno room in Detroit.)

Other modifications: the GM Turbo Hydra-Matic torque converter was unchanged except for raising the full-throttle shift point from 4600 to 4800 rpm. Also, they made a neat floor shift lever for Unser (though he stayed in "Drive" range most of the time), that permits up- and down-shifts at will. The overall gear ratio had to be changed substantially from the stock 3.21 ratio. This proved to be a job. Space requirements wouldn't permit a stiffer ratio than 4.09:1 on the spiral bevel final drive gears, using standard differential parts. So to get an overall ratio over 4-to-one, they simply used some old prototype chain sprockets that would give a step-up ratio of 4-to-one between the converter and gearbox. Thus, the overall ratio is 1.14 x 4.09—which works out to 4.67-to-one. This was calculated to give an engine speed in high gear of 5400 rpm at the maximum anticipated car speed of 100 mph. As it was, Bobby Unser got it up to a valve-clattering 107 mph at one point.

Special tires were a key factor in the success of this climb. Olds decided to use the highly effective 7.60-15 Goodyear Pikes Peak Specials. These are designed specifically for this course, and feature a deep, snow-type tread pattern with tiny pieces of walnut shells mixed in with the tread material. The idea here is that these sharp shell pieces will keep tearing out of the tread material, and thus keep the rubber surface rough and jagged. It's possible to develop substantial lateral *g* forces with these tires—almost like normal passenger tires on asphalt. Incidentally, the tread *width* proved to be a critical factor, even though the same tire size was used all the way. They ended up with 8-inch tread width on the front and 7 inches on the rear, mounted on special 8-inch rims that were welded to standard Toronado wheel spiders. Inflation pressures varied from 24 to 30 pounds.

Olds put metallic brake linings on the car, but braking proved to be relatively unimportant on the climb, and standard linings would have done nicely. However, a very important modification consisted of replacing the standard steering gear with an early prototype unit that gave an overall steering ratio of 12-to-one (vs. 17.8 standard). This let Unser wheel through the tight turns a lot easier. Power steering was retained (standard on Toronado), with the flow orifices opened up to give quicker power response on quick wheel turns. (It's easy to "beat" a power steering system when the ratio is speeded up.)

So how did the thing handle on gravel turns?

John Fitch, sports car expert who is a consultant for Oldsmobile, was on hand and waxed rather lyrical on the handling attributes: "Maybe we were 17 seconds off Jones' all-time stock car record, but it's also nice to remember that we were almost eight seconds *quicker* than Nick Sanborn's 1965 winning time of 14:17.7 in a racing Plymouth Hemi. The interesting thing about the Toronado's cornering power and handling qualities is indicated by separating the times taken on the fast straight sections from those on the tight curving sections of the climb. Although the Toronado lost time on the fast sections where cornering and handling matter less than a brute horsepower-to-weight ratio, it more than made up the deficit in the tight hairpins and slow bends."

Bobby Unser is a man of fewer words, but those words are right to the point. "That rascal fools you. You can get on the power a lot sooner and harder than you think coming out of the turns. But the front end pulls so effortlessly that it takes a few runs to really believe it's as good as it seems. I could hardly believe some of the clocked times on certain sections. But I'm beginning to think front-wheel-drive is the answer for this hill. If you could build front- and rear-drive cars with all other factors equal, there's no doubt in my mind that the front-drive would be faster overall."

And engineer Ted Louckes had some interesting technical comments on the car's hill performance. "We made all our money in the tight turns and switchbacks," says Ted. "These areas are only a small percentage of the total 12-mile course length, but our clocked times were as much as 20 per cent faster through these areas than all-out USAC racing jobs. It was hard to believe. Of course, we lost time on the longer, faster sections because we had 100 to 200 horses less than the stock cars. Parnelli Jones exceeded 120 mph at certain points where our top speed was just about a stand-off. But it kind of makes you wonder what the Toronado could do with 600 horses!

"I took a ride through some of those tight sections with Bobby. I was on the bare floor, tangled in the roll bar, hanging on for dear life. But it was an experience.

"I think a key factor in this superior cornering with front drive is that you can point your driving thrust wherever you want it by flicking the steering wheel. With rear drive the thrust is always in line with the axis of the car. This factor isn't so important on wide, fast turns where steering angles are small, but it's a key on tight turns on gravel. And our customers will find that it's important on ice and snow."

And so went the Olds Pikes Peak project. The operation itself was a success—but what's just as important, the engineers learned something besides how to get up a hill in a hurry. For one thing, they now have a little different view of the effect of spring stiffness and roll rate on the cornering of the Toronado. I think you'll see the ride softened up a bit, maybe even before the '67 model run begins—but without any big sacrifice in handling. Also, they admit that they're interested in an anti-roll bar on the rear, to reduce the understeer tendency that is inherent with front drive. They didn't need one at Pikes Peak, but you could feel the difference on pavement. Another cute trick they tried at the Peak was to use up to three degrees negative camber on front *and* rear wheels. This could be felt on dirt, and could certainly be felt on pavement. The car seemed to like an excess of negative camber on the rear. This could be a sign of things to come.

In fact, we might not have even seen the last of the Toronado on Pikes Peak. C/D

Seen from the rear the Toronado's clean lines stand out well.

THE TRUTH about the TORONADO

Impressions of the Revolutionary 7-litre Front-Wheel-Drive Oldsmobile

I COULD not resist using the above heading to this article, but in fact quite a number of people who were confronted by the Toronado were quite certain it couldn't be true and must be the property of Lady Penelope of International Rescue, a substitute, perhaps, for her well-known and formidable six-wheeled Rolls-Royce. . . .

In fact, this is unfair to the General Motors product which, apart from its very impressive size and the allure of being the only American front-wheel-drive automobile in normal, current production, is an eminently practical means of transport, and one which provides an extremely high degree of comfort and quietness. Although it is the most powerful front-drive car by a great many horsepower, in normal use it is virtually impossible to tell that this is so, but when accelerating hard or negotiating slippery roads it is definitely reassuring to be pulled along by those claimed 385 horses when treading hard on the accelerator and ambitious handling brings out the expected f.w.d. characteristics of understeer countered by nose-in cornering on a light or trailing throttle. The power-steering, although high geared at under 3½ turns lock-to-lock of a small, low-set steering wheel, in conjunction with poor turning circles, is feather-light and conveys absolutely no feel of any road existing beneath the driven wheels, although there is ample castor-return action. But the very effortlessness of steering control means that this huge motor car can be cornered fast and asked to change direction with impunity and suddenness; it is then that the impeccable directional control endowed by driving the front wheels is a definite asset.

The power for driving the Toronado's front wheels is provided by a special version of the 6,965-c.c. Oldsmobile Rocket V8 engine.

I confess I have only the half-truth about the Toronado because it would be necessary to be able to measure front tyre wear over several thousand miles and thrash it round a race circuit like the Nurburgring to know all, whereas I only drove it for a few hundred miles. But regarded as road transport it has absolutely no vices other than those found in most American automobiles and front-drive certainly enhances its fast cornering, acceleration and " dodgeability " stability.

The Oldsmobile Toronado, then, represents a genuine step forward by the great General Motors' empire; I am only disappointed to have to report that MOTOR SPORT had to wait until now to report this fact because the first Press car imported to Britain caught fire and was destroyed for no better reason than a fuel pipe chafing on a bulkhead, a defect not entirely isolated, and one which escaped the perception of the mighty G.M. research organisation. A spare length of fuel pipe with fitting instructions accompanied the test car, proof that emergency steps have been taken to rectify an alarming defect.

That apart, the 7-litre V8 Oldsmobile Toronado can hardly be faulted, for those who like this species of motoring. The silent running is superb, the action of the Turbo Hydra-Matic 3-speed automatic torque converter transmission extremely smooth. The performance, if not exactly sporting, is the equal of the better, bigger American automobiles, and there is full automation, for windows, their rear ¼-lights, radio aerial and all positions of the seats—the seat adjustment is like that of the Mercedes-Benz 600 but done electrically instead of hydraulically. The headlamps are normally recessed beneath flaps, and if you can afford the extras there is very thorough ventilation, with refrigerated cooling. All of which adds up to the mostest up-to-dateness!

There has been no attempt to exploit the compact possibilities of front-wheel-drive, possibly because American customers buying a 425 cu. in. car in this price range expect plenty of automobile for their $6,000, and also because a good deal of space is inevitably going to be occupied by a 385-h.p. V8 engine and automatic gearbox, whether it drives the front or the back wheels.

General Motors have, however, eliminated the transmission hump and drive-line tunnel and have adopted the ingenious method of mounting the engine slightly to the o/s, driving the torque converter normally from the rear of it, but turning the rest of the gearbox round, so that it can be accommodated beside the engine on the n/s, facing forward, the drive being taken across from converter to transmission by a rubber-dampened link chain. A differential unit is bolted to the front of the gearbox and drives the front wheels through Saginaw universal joints, consisting of ball-spline Rzeppa inboard and grease-prepacked Rzeppa outboard joints. The Toronado runs on low profile 8.85 × 15 tyres. It is available at present in one version only, a Fisher 2-door, 6-seater hard-top coupé, in standard and de luxe forms. It is aggressive-looking from the front, quite plain seen from behind and most of the flamboyance of this revolutionary Oldsmobile stems from size alone—although the wheelbase is only 9 ft. 11 in., the track is 5 ft. 3½ in. in front, 5 ft. 3 in. at the back, and the Toronado has an overall length and width, respectively, of 17 ft. 7 in. and 6 ft. 7 in. It is fashionably low, at 4 ft. 6¼ in. unladen.

The finger-tip controls for windows and front seat adjustment in the driver's door of the f.w.d. Oldsmobile.

The house in the background is where Col. Cody lived while conducting his pioneer flying experiments on Laffam's Plain, Farnborough. It is of interest in this 100th anniversary year of the Royal Aeronautical Society. The Toronado is faster than Cody's aeroplanes and he could hardly have foreseen such potent road vehicles!

There is no excuse for not sitting comfortably in this gigantic Toronado, because in addition to electric seat selection, a small l.h. lever under the steering wheel enables the steering wheel, with its dual spokes that incorporate the horn-pushes, to be tilted as well as raised or lowered over a wide range. And although the instrumentation gives the impression at a casual inspection of being second-cousin to a small aeroplane, it is, in fact, logically laid-out and well-contrived. A horizontal drum-type speedometer, rotating downwards to 130 m.p.h., is a bit distracting but has large, clear figures and is notably steady-reading. The push-buttons for 2-speed wipers and washers are to hand on the left of the hooded cluster, and the engine starts—always immediately because the automatic choke functions splendidly—on the key if the gear selector is in N or PARK.

The gear selector is normally left in D, but can be moved freely into S, which gives hold control for better acceleration, while the normal kick-down is naturally incorporated. There is also the usual L position. The only dials are those for fuel contents, water temperature (calibrated C-H) and ammeter, the considerable charge reading on the latter a reminder of the extent of the electrical services. A red indicator is a warning that the foot-operated parking brake has not been released, and green lights flash when the winkers, controlled by a precise left-hand stalk above the steering-wheel adjuster lever, are in use. Otherwise, the apparently complicated controls resolve themselves into air-conditioner settings, rear window de-froster settings, interior lamp switch, and speaking/bass/music selection, and front/rear speaker selection for the Oldsmobile transistor radio. Other anticipated warning lights are there, the speedometer incorporates no trip mileometer but the total odometer reads to tenths, there are air vents each end of the facia and ash-tray and cubby-hole on its wall, cigar lighters are provided, and the switches for the seat and window electrical operations are on the driver's door sill, duplicated for window control on the passenger's door-sill. The windows rise and fall rather noisily but efficiently; the doors, which have sill interior locks, are rather heavy and springy to shut. The interior lift-up door handles are duplicated to serve both front- and rear-seat occupants.

The interior decor of the Toronado is typically trans-Atlantic but not really garish. Black plastic upholstery gives a neat, restful appearance to the comfortable seats and there is plating of control stalks, switches, etc., but otherwise not any fancy gimmickry. The external appearance is enhanced by the cut-away of the road wheels to provide cooling for the finned brake drums.

Coat hooks, seat-belts for back as well as front-seat passengers, white and red lights on the trailing edges of the doors, lots of back-seat leg room, less head room, a very big vanity mirror in the n/s vizor, easy access to back seat in spite of only two doors, a beige colour finish, central arm-rests on the front seat, are aspects of the Toronado appreciated by passengers as well as drivers.

Bonnet and boot lid are self-supporting and the spare wheel and lifting jack are mounted on the front wall of the big boot. There is illumination of the drop-lid cubby-hole, which is the only oddments stowage provided, and for a huge ash-tray. The external mirrors were well set and much appreciated on this ultra-wide vehicle, and the driver's door carried an additional external mirror. The screen wiper blades work in opposite directions, the

Instrumentation in the Oldsmobile Toronado is not so complicated as it appears at first sight. These pictures show the tilting adjustment of the steering wheel.

With its aggressive frontal aspect and concealed headlamps the Oldsmobile Toronado is not easily forgotten by those who meet it coming in the opposite direction, especially on narrow English roads.

test car had British flasher-units-cum-side lamps, the doors incorporated long narrow sills or arm-rests but lacked pockets and the rear seats were shaped to some extent, as if for two occupants. The ignition had to be on before the electric window lifts would work. An amusing minor detail was the slogan "Your key to greater value—G.M." on the Briggs & Stratton door keys, made in Milwaukee.

So much for detail impressions. There is not much to add about driving impressions, so easy to control, so effortless in operation is this two-pedal Oldsmobile Toronado. It has all the performance one is likely to crave up to a maximum speed of 128-130 m.p.h., steers accurately (with one finger!), corners far "flatter" than most American cars but has the usual inefficient brakes, prone to severe fade if used hard from high speeds, although just about adequate, but snatchy, otherwise. This 2½-ton monster plows over bad roads rather than rides them, so comfort is good and with the optional automation it represents lazy, efficient travel and impresses as a fine piece of automotive engineering.

The dual Guide sealed beam headlamps take about 4-6 seconds to "unflap," so are useless for emergency flashing, nor is any separate flasher switch provided. These, and the side-lamps, are switched on from a knob convenient to the left hand of this l.h.d car, and rotating the knob controls instrument lighting, as on a Vauxhall. The mirror is anti-dazzle and corner illumination comes on with the flashers.

The sheer width of the Toronado is rather intimidating on many British roads, especially as the arches over the wheels protrude, adding to overall width. The ventilation system, with venting under the back seat and from louvres under the back window, has obviously been very thoroughly contrived.

The Toronado uses a boxed chassis frame, torsion-bar front suspension and a back axle of U-shaped channel carrying dead spindles, sprung on two single-leaf springs and damped by four shock-absorbers—the Toronado claims to be the only car with four rear shock-absorbers.

The engine is a modified, more powerful version of the 6,965-c.c. Oldsmobile Rocket V8, developed by better breathing, a new 4-barrel Rochester Quadrajet carburretter, with 22% smaller primary but 44% larger secondary fuel stages, larger inlet valves and a high-lift camshaft and larger tappets.

Here I must remark on the very detailed data issued by G.M.'s Press department. Although readers may feel that how we obtain the information we pass on to them is no concern of theirs, an efficient Press service usually reflects engineering and sales efficiency in the company concerned—and General Motors' Press liaison is certainly thorough. I suppose, however, that those who, rightly or wrongly, have been hysterically condemning General Motors and other American manufacturers for callously building accident-prone cars will not agree that this could extend to safety research!

From driving out of London on a full 20-gallon tank of

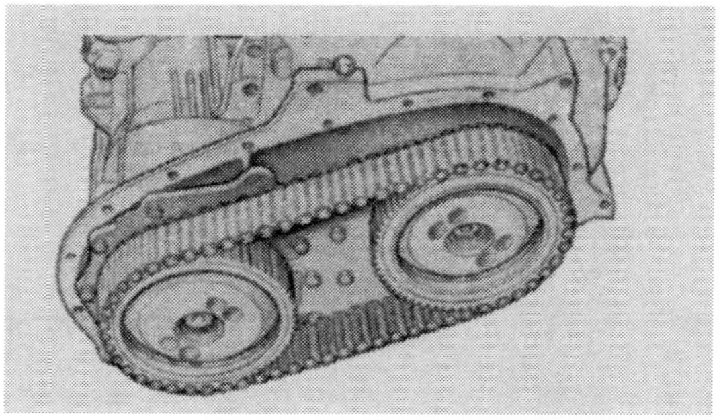

TORONADO UTILIZED POWER PACKAGE.—The transmission is split with the torque converter connected to the engine crankshaft and the mechanical transmission connected to the differential. A chain and sprocket drive arrangement transfers engine power from the converter to the transmission.

premium petrol the fuel gauge got close to empty after 185 miles. Overall consumption was 10.8 m.p.g. and sump level after driving 700 miles was restored with a pint of oil. Chassis lubrication is called for only at 12,000-36,000-mile intervals. But as the de luxe car, with extras as tested, costs £4,415 12s. 1d. inclusive of import duty and purchase tax, petrol thirst, which really isn't excessive, is unlikely to trouble the Toronado owner. I rate this 7-litre front-wheel-drive Oldsmobile as one of America's best cars. It is covetable in certain quarters as the most up-to-date of its kind but it also merits consideration in the same quarters as a very fast, quiet, safe-handling, fully-automated and air-conditioned modern automobile, exceedingly well engineered. I must be queer, for I preferred driving a vintage Sunbeam and handed the Oldsmobile over to the photographer half way through the test. Fortunately for General Motors, not everyone feels that way!—W. B.

THE OLDSMOBILE TORONADO 2-DOOR COUPE

Engine: Eight cylinders in vee formation. 104.8 × 100.9 mm. (6,965 c.c.). Push-rod-operated overhead valves. 10.5 to 1 c.r. 385 b.h.p. at 4,800 r.p.m.
Gear ratios: Low, 7.96—17.49; Intermediate, 4.75—10.46; Top, 3.21—7.06 to 1. Automatic transmission.
Tyres: 8.85 × 15 Firestone 500 low-profile nylon semi-whitewall, on ventilated bolt-on disc wheels.
Weight: 2 tons 6 cwt. (kerb weight).
Steering ratio: 3.4 turns, lock-to-lock (power steering).
Fuel capacity: 20 gallons. (Range: approx. 200 miles.)
Wheelbase: 9 ft. 11 in.
Track: Front, 5 ft. 3½ in.; rear, 5 ft. 3 in.
Dimensions: 17 ft. 7 in. × 6 ft. 6½ in. × 4 ft. 6¼ in. (high).
Price: £3,365 (£4,067 12s. 1d. inclusive of purchase tax). With extras as tested: £4,415 12s. 1d.
Makers: Oldsmobile Division of General Motors Corporation, Lansing, Michigan, America.
Concessionaires: Lendrum & Hartman Ltd., Flood Street, London, S.W.3.

Performance Data

Acceleration:

0-50 m.p.h.	5.7 seconds
0-60 m.p.h.	12.5 seconds
0-100 m.p.h.	19.7 seconds
s.s. ¼-mile	16.4 seconds (93 m.p.h. over finishing line)
s.s. ½-mile	26.0 seconds (110 m.p.h. over finishing line)

TORONADO: THE BIG WIND

IT'S AN ILL BREEZE THAT BLOWS NO GOOD, AND THE TORONADO WILL BLOW YOU OFF QUICKER THAN MOST!

From SLONIGER, far from his own stamping grounds.

WHAT more appropriate place for a second generation Oldsmobile Toronado than the redwoods of California, often billed as the largest living things in our world. Even in western America where eight lanes are a normal city two-decker strip, parking spaces look truck sized to Europeans and the scenery scales a Corvair down to an NSU — the Toronado seems vast on first acquaintance from its beetle-browed driving seat.

This mass fading off toward the horizon is deceptive, you know. In fact, this pacesetter is smaller in both wheelbase and overall length than the normal big Olds — the Ninety-Eight. The fastback really falls between the compact and a full-size kind in all dimensions except power and weight. With 385 bhp the front-wheel Toronado is the most Olds engine you can buy (naturally there are bigger powerpacks from other companies, Olds is part of the GM no-power syndrome). At 4475 kerb lbs it is also the heaviest Olds by a hair.

It is most likely the thirstiest as well although I never got a wholly satisfactory fuel figure of consistency in nearly 700 miles. Perhaps the car was just loosening up but we started at 13 mpg (imp) and ended at 15.3 with an average of 14.6. Yet all the runs were more or less equal mixtures of freeway, two-lane corkscrews and city action — if San Francisco's vertical scenery can be called city.

Quite obviously the fact that this car is pulled along rather than pushed like all its peers bar the new Eldorado, should lead to a very different kind of handling. After all the Mini-man once commented that 2-litres was maximum for a usable FWD package. The Toronado has 7-litres up front — though of course Issigonis might point out he meant workable in

EVERYTHING'S big in yankee-land, but even the General's giant Toronado can't hope to compete with the Californian redwoods.

Sloniger's just-to-prove-I-was-there shot of the Toronado to a background of the famous San Francisco Golden Gate bridge.

the Makinen style of Mini motoring. But I doubt if any US driver would even know the big Olds had its traction at a different end than his other American iron. All US cars are understeerers (nose-heavy). This is just more of the same.

And in a land without 65 speed limits and stately flow? I wouldn't really want to storm the Alps in this 110 mph automobile but at the US speeds which control its drum brake limits the quirks are pure American, not FWD. You won't be transferring any Citroen driving habits to the Toronado.

Brakes are the ruling fact of Toronado life. With discs the in thing here now it seems inconceivable that Oldsmobile put all drums in their toy. The local argument goes that US drivers never check a car unless it stops dead in front of a topless shoeshine girl at high noon

Two by the seas: Sloniger imported his own bird (wife) for his US jaunt, but relied on local transport. Olds got him from coast to coast.

and therefore would weld the puckless pistons to the discs in six months. Perhaps. But this speciality car is supposed to attract different drivers according to the ads. It has automatic outside mirror adjustment, would a brake wear sensor be too much to ask?

Anyway the big bull (Toro might be a very handy nickname for this one) uses four drums, finned I grant you and stuck behind fretwork wheels for cooling. They even give the beast the largest wheels going at Oldsmobile — full 15 inchers mounting 8.85 rubber — but it is flatly underbraked for 2¼ tons of car and people.

Speaking of people, Oldsmobile bills this one as a six-place car but gives it the lowest front headroom and least rear legroom of the 1967 line. At very roughly 5500 of your dollars it strikes me as the epitome of the personal automobile for two wealthy people who take an occasional friend along when his Caddy is in for service.

GM got more useable room than cars of similar dimensions of course by eliminating the hump, front and rear. Despite the vast doors — "new for '67" — and inadequately hinged to close properly — the minimal roofline makes it easier to load three in front and throw your minks on the rear seat. Incidentally seat back latches have reached GM (wisely) and the Toronado has them extremely well sited.

In keeping with its waft along image the Toronado has power everything, with brake and steering booster standard, the power windows and seats an option. Optional extras are the name of the game in American car buying. Buyers think nothing of $1000 on top of a $3000 automobile.

The game we enjoyed most was power-adjusted seating. Three little toggles let you go up and down, back and forth and tilt in four planes at will. That sudden recline feature should give bachelor owners all sorts of fun. More practically, when combined with a steering wheel which goes in and out on its (new this year) GM collapsible column and tilts up and down for rake adjustment, the combination makes it possible to drive an Olds with almost European stretch.

The wheel really has far more tilt range than anybody will use but it makes a pleasant gimmick to amuse the kiddies. Since the heavy automobile needs power steering as much as any made, because it has that FWD and massive doughnuts to turn, GM took advantage of the lowering process to fit a miniscule steering wheel. I'd like another inch on the overall height and a wheel slightly larger than my late lamented kart.

While we are on the subject of items GM can leave off my Toronado in its third year — drum speedometers don't turn me on any more than drum brakes. It may look pretty and have nice large numerals but it takes study rather than a flick glance. Blind rear quarters are probably inherent in the fastback mode but while I like tinted windscreens, I would like a little more of them vertically.

As for passenger comfort, the heater/demister in front worked immediately and with some marked degree of temperature control but that expanse of rear glazing could do well with a blower all its own. Wind wings were eliminated, and are unneeded with their through-flow ventilation but even an Oldsmobile will fog up inside.

The Toronado's finest hour comes on those long, open highways where it purrs along gently and effortlessly, never interrupting the stereo radio. Olds boasts of a softer, quieter ride for 67 with lower spring rates and new shock absorbers, two in front, four in back. These and the body mounts seem to have been the chief engineering changes to go with very minor alterations in grille and tail lights. They underline the fact that the debutante Toronado was pretty close to expectations.

Now that the Toronado has settled into the scenery virtually nobody even turns around to look and I'll wager half of America doesn't know this Olds has front-wheel drive. Most likely even less care. Moving from novelty to everyday — if expensive — transport the second-generation Toronado gives every promise of a long life.

STORY BY JOHN ETHERIDGE   PHOTOGRAPHY BY BOB D'OLIVO

5 LUXURY SPECIALTY CARS

☐ Before the advent of the Luxury Specialty car (T-bird was first), the buyer who wanted something with more spice than the stolid domestic cars could offer, even in their most jazzed-up forms, had to go to something from overseas. These classy imports had strong esthetic appeal and stood out like Brigitte Bardot in a line of Twiggys at a bikini show. ☐ But all was not wine and roses with these machines because their creators hadn't taken conditions in this country into account. It was always with great trepidation that the luxury import owner left his pride and joy stand on an urban street or parking lot, because likely as not some nice old lady, parking with all the tenderness and care of a destroyer ramming a U-boat, would de-flower it before his errand was accomplished. ☐ Besides having to replace and straighten flimsy bumpers and sheetmetal, frequent trips to the repair shop were necessary because these cars were very finicky compared to most of the domestics. And all too often garage owners, assuming anyone who would drive such a car to be rich and foolish — and hence fair game, turned into Black Barts in shop coats and indulged in outright banditry. ☐ In addition to this group whose contretemps with foreign thoroughbreds in the '50s and '60s reverted them toward Detroit iron, there was a much larger number with esthetic leanings but who were too firmly attached to the American way of motoring to consider an import. Walk-in doors, flower vases, lap warmers, and general stodginess of the home product were not qualities that endeared a car to a member of the latter group. It was things like reliability and durability, power accessories, air conditioning, and automatic transmissions that kept them buying American. ☐ The Luxury Specialty car that came to fulfill this market could, with a great deal of truth, be said to be the kind of car everyone would own provided it were cheap enough and there were no special space or use requirements. Motorists of just about any stripe can find among our test group a *now* car with pleasing and distinctive lines, good performance, and all the things that go to make a car enjoyable. ☐ The erstwhile exotic import fancier may give up a few mpg and, perhaps, some maneuverability for something otherwise much easier to live with. Likewise, the big luxury car driver will have to do with smaller luggage space and comfortable seating for four or five instead of five or six in some cases. But neither has to make any painful adjustments, and both readily agree that the Luxury Specialty car is an excellent compromise. ☐ The five cars in this class compete with one another in a sense but nothing like the head-on, feature-for-feature struggle we find in the Low-priced Specialty cars. Each has its own individuality or personality, so to speak. This holds true in spite of the fact that three of them — the Toronado, Riviera and Eldorado — share the same basic body shell and two — the Toronado and Eldorado — have almost identical fwd and suspension systems. ☐ Also there's a large price differential, depending on what optional equipment is ordered between the Grand Prix at the lower end and the Eldorado on top. So, strictly speaking, all don't compete pricewise, either. But this doesn't mean that one prospective buyer wouldn't consider all five before narrowing down his choice. In this class more than any other, buyers will unhesitatingly go up or down several

5 LUXURY SPECIALTY CARS

clockwise
from top left:
eldorado
thunderbird
toronado
riviera
grand prix
eldorado

thousand dollars to get the styling or some particular feature they want.

POWERTRAIN & PERFORMANCE

Since the Thunderbird and Grand Prix offer optional engines, we chose engines for them that were the nearest equivalent to sole engine offerings of the Toronado, Eldorado and Riviera. The resulting engine line-up is pretty evenly matched — within 5-cu.-in. displacement and 45 advertised hp. The Grand Prix, which comes standard with a 3-speed manual transmission and offers a 4-speed manual as an option, was ordered with the optional 3-speed automatic because all of the others have this type with no options.

The Riviera GS is the outstanding performer of the group, tested chiefly because of its 3.42 "performance" axle ratio. The standard 3.07 ratio produces performance more in line with that of the others. The Riviera lost none of its quietness, and the engine remained unobtrusive as ever in spite of the considerably higher numerical ratio. Undoubtedly the Star Performer would be a Grand Prix with the 428 HO engine (376 hp @ 5100 rpm). Maybe a GP with this optional romping, stomping powerplant and an automatic would still be tame enough to be considered in the Luxury Specialty class, but it definitely would be something else with a 4-speed and heavy clutch.

As can be seen from the spec table, all cars are highly satisfactory performers, although performance *per se* is not an overriding factor in this kind of car. Thunderbirds with the smaller 390, and GPs with the 400 engine, give somewhat improved economy and performance that is still more than adequate. With any optional or standard engine on any of the cars, relaxed cruising rpm with plenty of reserve power — both for accessories and passing — is the order of the day.

We almost forgot to mention that the Toronado and Eldorado do their driving with the front wheels. And that's a good indication of how noticeable it is. Only under unusual conditions like cornering much faster than normal do you feel anything different. Also, we noticed the fwd cars would move out smartly from a standstill on rain-drenched streets, while their conventional brethren tended to lag behind with churning rear wheels.

There was nothing in our experience with Ford's automatic transmission in the Thunderbird and GM's Turbo Hydra-Matic in the other cars to make us prefer one over the other. Each went about its business with quick, barely detectable shifts.

HANDLING, STEERING & STOPPING

With the exception of Thunderbird all cars in this group either come with stiffer-than-usual suspensions (Toronado and Eldorado) or offer some option in this department (Riviera's GS package and the GP's Ride and Handling package). Whether loyal T-birders would go for some sort of handling kit we can't say, but the car suffers in comparison to any of the others in this respect. It is *very* softly sprung and has almost no roll stiffness.

But such a suspension does have its virtues. At city speeds it smoothed out potholes, railroad crossings, etc., like none of the others. The Toronado and Eldorado suspensions tended to show their teeth under similar conditions.

The Grand Prix with standard suspension had an excellent ride, yet was stable and had reasonable steering response. To our way of thinking, the handling kit would make it more fun to drive, but the car is quite acceptable as is.

We can't decide whether the Riviera GS or Toronado has the best ride/handling combination, but the two of them are clearly ahead of the others in this regard. Both are truly superb road cars that beg to be cruised around 100 mph. They feel very secure and stable at high speeds and are practically immune to crosswinds, undulating surfaces, and other perturbations.

The Eldorado's suspension differs from the Toronado's in that it has lower rate (less stiff) rear springs paired with a load-leveler as standard equipment. Thus riding height and natural frequency stay pretty much constant with varying loads. Because tires are as much a part of the suspension as the springs, there's another difference. The Toronado uses a special low-profile tire while the Eldorado uses the same tire as the rest of the Cadillacs.

The Toronado tire has a wider tread, and the low section height is more stable. Hence this tire puts more rubber on the road and keeps it there. The Eldorado has noticeably less cornering power and all-around traction than the Toronado for this reason. Disc brakes are practically a necessity on the Eldorado as stops from 60 mph in less than 200 feet are hard to come by with drums and these tires. The drum-braked Toronado stopped in about the same distance as the disc-braked Eldorado due to the difference in tires. From all indications a Toronado with discs should be a super stopper. Also, the rear wheels on the Eldorado had an annoying tendency to lock very early and cause fish-tailing when braking on wet surfaces, and again we can think of no reason except the tires.

The standard Thunderbird disc/drum combination is very well balanced and produces the ultimate in stopping power. What we've said that discs could be expected to do for the Toronado applies to some extent to the Grand Prix and Riviera, too: a further improvement in stopping, even though performance of their drums was very creditable.

SPACE, COMFORT & SAFETY

Space and comfort for a stated number of passengers is practically the name of the game with this class of car, and taken as a group they are probably safer than the average car on the road today. They are very kind to occupants — especially the driver — on extended trips, and it's pretty generally conceded that the fresh, alert driver is a safer driver.

Either fastback design and/or smallish rear windows tended to limit rearward visibility on all except the Grand Prix. But none was, in our opinion, restricted to the point of being hazardous, and with use of outside mirrors and the pivots in our neck, we had no difficulty determining when the coast was clear for lane changing.

BEST & WORST FEATURES

Trying to be objective about these cars proves to be as elusive as trying to do the same with another class of beguiling creatures — women. In looking back on our encounters with them we tend to remember not so much which was best or worst, just that in some ways some were better than others.

The big doors of the Riviera, Toronado and Eldorado qualify for both categories. They give good access to the rear seat, but need lots of room to swing open.

The best feature of the Eldorado, which has many nice accessories including a fine AM-FM Multiplex Stereo, has nothing to do with any of these. It's the awesome respect and prestige value accorded it by persons from all walks of life. After observing the effect it had on a great many onlookers we're convinced that, in this country at least, there's no car made anywhere at any price that equals it in this respect. Its worst features are the tires and the bent rear window that puts a sinister sneer on the grille of any car approaching from behind.

The T-bird's flabby suspension is its biggest drawback, and the good brakes and availability of a smart appearing 4-door are the best things it has going for it. The Grand Prix is the only one of the bunch offering a full choice of engine and driveline options as well as a convertible body style.

Flow-through ventilation (which we're sold on) and absence of front vent panes was a feature on all five cars. The missing vents and placement of the door-locking buttons on some of the cars probably won't stop a determined thief, but it will tend to separate the professionals from the joy-riders.

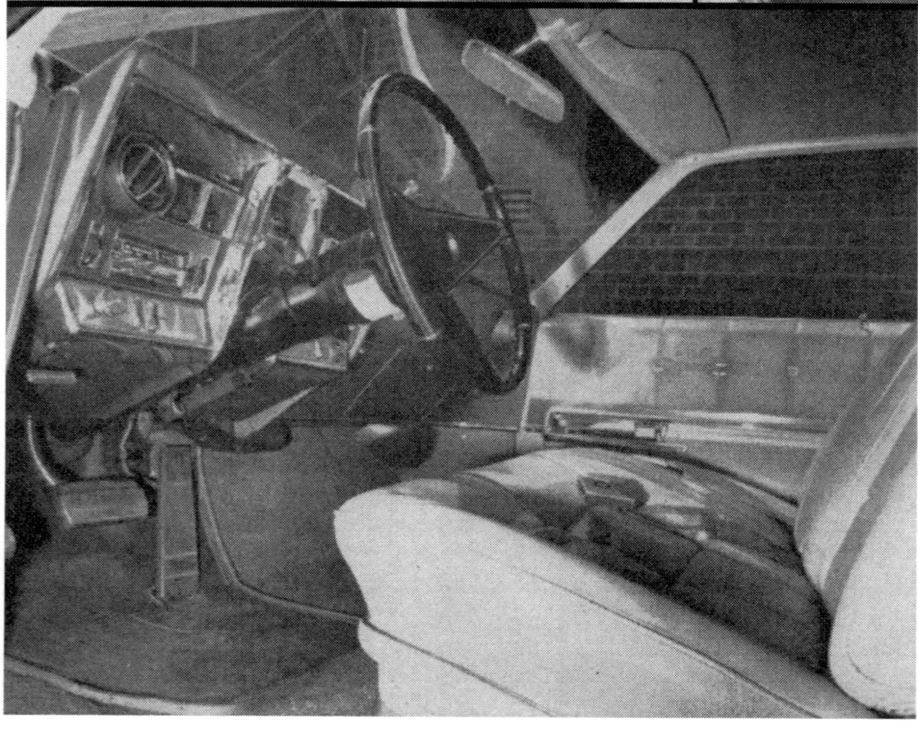

ELDORADO — GRAND PRIX
THUNDERBIRD — RIVIERA
TORONADO

Leather buckets are optional on Eldorado but bench shown here has fold-down center armrest, serves same function and has proved most popular. Grand Prix, shown with standard buckets, offers bench option. T-bird 4-door brings you luxurious expanded-vinyl tufted fabric, vinyl trim. Vinyl-interiored Riviera comes standard with tilt wheel — a needed option on others for complete comfort. Toronado in leather-textured vinyl has neat, designed-for-this-car look about its interior furnishings.

PHOTOS BY BOB D'OLIVO, GERRY STILES

	Eldorado	Grand Prix	Riviera GS	4-Door Thunderbird	Deluxe Toronado
PERFORMANCE					
Acceleration (2-aboard)					
0-60 mph (secs.)	9.5	8.4	7.8	9.0	8.9
1/4-mile (secs.)	17.0	16.1	15.9	16.8	16.6
Speed at end of 1/4-mile (mph)	81	87	86	86	85
Mph per 1000 rpm	23.7	27.0	22.6	25.0	25.3
Stopping Distances					
From 30 mph (ft.)	41	40	38	40	38
From 60 mph (ft.)	165	167	165	143	168
Speedometer Error (%)	+3	+2	0	0	+3
Gas Mileage Range (mpg)	10-13	10.5-14.5	10-13.5	10-13.5	10-13
SPECIFICATIONS					
Engine Type	V-8	V-8	V-8	V-8	V-8
Bore & Stroke (ins.)	4.13x4.00	4.12x4.00	4.19x3.90	4.13x3.98	4.12x3.97
Displacement (cu. ins.)	429	428	430	428	425
Horsepower @ rpm	340@4600	360@4600	360@5000	345@4600	385@4800
Torque (lbs.-ft. @ rpm)	480@3000	472@3200	475@3200	462@2800	480@3200
Compression Ratio	10.5:1	10.5:1	10.25:1	10.5:1	10.5:1
Carburetion	1 4-bbl	1 4-bbl	1 4-bbl	1 4-bbl	1 4-bbl
Transmission Type	3-spd. Auto	3-spd. Auto	3-spd. Auto	3-spd. Auto	3-spd. Auto
Final Drive Ratio	3.21	2.93	3.42	3.00	3.21
Steering					
Type	Variable-Ratio Power	Power	Power	Power	Power
Turning Dia. Curb-to-Curb (ft.)	41.3	42.8	42.3	42	43
Turns Lock-to-Lock	2.6	4.2	3.57	3.68	3.4
Wheel Size	15x6JK	14x6JK	15x6L	15x5.5JK	15x6JK
Tire Size	9.00x15	8.55x14	8.45x15	8.15x15	8.85x15
Brakes	Opt. Disc/Drum	Drum	Drum	Std. Disc/Drum	Drum
Fuel Capacity (gals.)	24	26.5	25	24.1	24
Usable Trunk Capacity (cu. ft.)	13.46	19.4	10.32	12.3	14.1
Curb Wt. (lbs.)	4680	4400	4420	4640	4800
Dimensions					
Wheelbase (ins.)	120.0	121.0	119.0	117.0	119.0
Front Track (ins.)	63.5	63.0	63.5	62.0	63.5
Rear Track (ins.)	63.0	64.0	63.0	62.0	63.0
Length (ins.)	221	215.6	211.3	209.4	211.0
Width (ins.)	79.9	79.4	79.4	77.3	78.5
Height (ins.)	53.8	54.2	53.2	53.8	52.8
PRICES AND ACCESSORIES					
Manufacturer's suggested retail	$6277.00	$3549.00	$4791.88	$4858.25	$4869.00
Optional Engine	—	(428 V-8) 78.99	—	(428 V-8) 90.68	—
Air Conditioning	515.75	419.60	421.00	421.49	421.28
Automatic Transmission	Std.	226.44	Std.	Std.	Std.
AM Radio	161.60	82.25	NA	NA	86.89
AM-FM Radio	187.90	124.20	175.24	89.94	173.78
AM-FM Stereo	287.90	225.00	266.81	163.77	238.03
Stereo Tape	NA	116.00	115.00	128.49	128.49
Vinyl Roof	131.60	105.32	115.78	Std.	110.59
Power Windows	Std.	104.00	105.25	103.95	104.00
Power Seat	83.15	94.79	94.73	97.32	94.79
Power Steering	Std.	105.25	Std.	Std.	Std.
Disc Brakes	105.25	110.50	78.94	Std.	78.99
Speed Control	94.75	44.95	63.15	129.55	84.26
Tinted Glass	50.55	42.10	42.10	47.49	47.39

NEW CADILLAC AND OLDSMOBILE

There are few external changes to the front-wheel-drive Cadillac Eldorado but a completely new 7.7-litre engine under the hood

FOLLOWING the Buick and Pontiac styling and engineering changes for 1968, General Motors have now released details for Oldsmobile and Cadillac. Chevrolet's new models will be announced next week.

The main story from both the prestige (Cadillac) end and the popular (Oldsmobile) ranges concerns engines. Cadillac have increased the capacity of their vee-8 from 7,031 c.c. to 7,736 c.c., making it the biggest unit in production in the world for passenger cars. Maximum gross power is up from 340 to 375 b.h.p. and torque is increased from 480 to 525 lb.ft.

Every part of the engine except four small components is new and it has taken three years and over two million test miles to develop it for production. To comply with the new Federal controls on pollution the new unit has an integral distribution system for air injection into the exhaust. Mountings for air conditioning equipment are also built into the structural castings for the first time. Last year 88.4 per cent of Cadillacs were ordered with air conditioning.

Cadillac buyers expect a few gimmicks and the latest is a buzzer which sounds in addition to a red tell-tale lighting if the engine temperature rises for any reason.

Much of the construction of the new engine has been simplified to reduce the number of joints and gaskets, and the bottom end has been generally beefed up with stiffer crank and a 24 per cent increase in bearing surface area. The camshaft has been designed by computer.

Unlike Rolls-Royce and Mercedes, who compete most closely with Cadillac in the large, luxury car markets of the world, Cadillac production is high and last year nearly 200,000 were built of which some 72 per cent were the DeVille models.

Externally the new Cadillacs have only small changes to grilles and lamp clusters, with new internal fittings on the general safety theme. Wipers park out of sight and there are the new GM side marker lamps on each front wing.

OLDSMOBILE

There has been an integrated design policy carried out by Oldsmobile for two years to design a complete power train from engine to wheels. The objectives have been better economy, more usable performance, less noise and vibration and a reduction in exhaust emissions. The way to this end has been to raise overall gearing and design the engine to rev less. To gain performance, engine size has been increased from 6.9 to 7.5 litres and final drive ratios are raised from 3.08 to 2.56 to 1. This cuts the revs at 90 m.p.h. from 3,400 to 2,840 r.p.m. and at 30 m.p.h. the 1968 model is revving at only 950 r.p.m.

In order to make this reduction a special "low duration" camshaft has been designed to bring the peaks for the torque and power curves down a few hundred r.p.m. This plus the capacity rise brings the values down from 365 at 4,800 to 365 at 4,600 for b.h.p. and from 470 at 3,200 to 510 at 3,000 for the torque in lb.ft.

To give better acceleration as well, the torque converter ratio of the automatic gearbox has been raised from 1.8 to 2.3 to 1. This puts the maximum bottom gear ratio up from 13.7 to 14.6 to 1 despite the change in final drive.

To reduce exhaust emissions Oldsmobile are using their "climatic combustion control", with pre-heated air and careful carburation and ignition regulation. This was introduced as an option last year.

Other changes to Oldsmobiles are of a styling nature with a new front for the Toronado and some major revisions to the Vista Vision estate car. ■

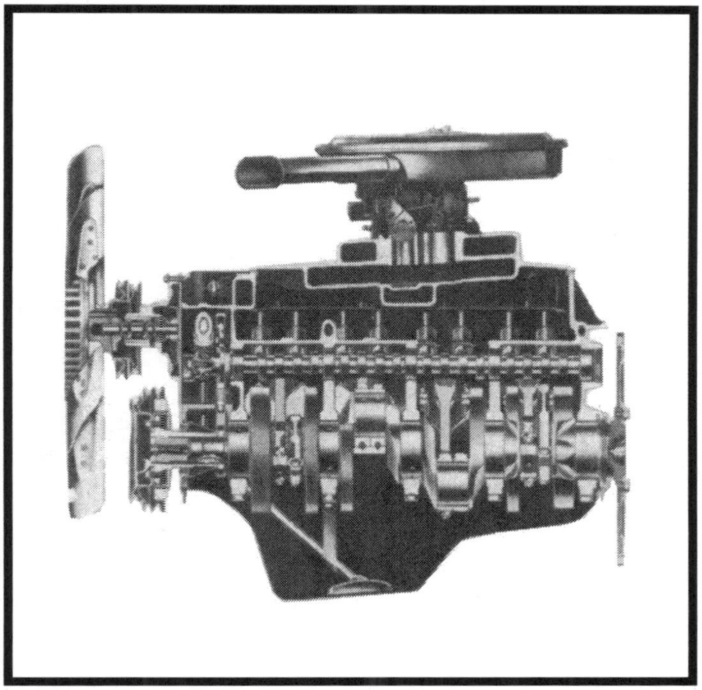

Top right: This is the high-performance Oldsmobile 4-4-2 with restyled body and side stripes. Above: Section through the new Cadillac engine showing the stiff crank and generous bearings. Below: Striking new front for the Oldsmobile Toronado

CAR and DRIVER ROAD TEST

Oldsmobile Toronado

A giant 400 horsepower, two-and-a-half-ton Mini is a gas, but it's not for citizens with cardiac conditions.

Luxury isn't enough in the luxury car field anymore.

It simply won't do to say, "My luxury car has more trick stuff than your luxury car," and so Oldsmobile's Toronado with its manufactured, Spanish-sounding name should be the last word in the neighborhood one-upmanship contest. After all, snow tires on the front have to be worth at least three points in establishing you as the Jones everyone has to keep up with.

The Toronado mystique—which began by being highly touted, found itself buried beneath comparisons to ordinary, everyday cars that we all know and love, and is now just barely becoming touted again—centers around its front-wheel-drive system.

One of Oldsmobile's problems in toutistry of fwd is that it's great to blather about, but it doesn't seem to be generally understood. The advantages of locating the engine and driving wheels on the same end of the car are fairly obvious. More weight on the driving wheels for traction and the absence of driveline tunnels intruding on passenger compartment space are desirable features in any automobile.

The problem then is simply choosing the right end. Engine in the rear presents the least mechanical problems since the driven wheels will not be required to steer. But, distinct stability advantages result from locating the weight forward. A car in motion is very much like an arrow in flight. Imagine a tail-heavy arrow or, to be closer to the subject, a VW in a crosswind. Since the arrow analogy establishes a better than good case for fwd, will there be any disadvantages other than cost? Sadly, yes. Whenever the most heavily loaded pair of tires is also the *driven* set there will be a certain reduction in ultimate cornering ability. This applies only to passenger cars where the same size wheels and tires are used both front and rear.

Now in fwd cars, front tires are expected not only to carry more than their share of the load and supply all of the tractive force, but also to do the steering. As a result, the cars normally have peculiar handling characteristics and extraordinary front tire wear.

Knowing the principles, it comes as no surprise to find the Toronado performing exactly as expected.

The popular fwd cars in the world: the BMC Mini, the MG 1275, and the Renault R-16, are economy cars designed to make the best use of available space. Their box-like bodies allow maximum usable passenger space, they have flat floors and a rear suspension design that does not intrude on luggage space. Not so the Toronado. To carve a notch out of the Thunderbird/Riviera market, Oldsmobile has chosen a unique approach. Attractive styling—probably the most important single commodity in this price class—precludes the use of the boxy shape. And right away, there goes one of the most significant advantages of the fwd concept: large passenger and luggage space. Happily for Oldsmobile, that in itself is not a serious loss in the Toronado's marketing slot. On the other hand, what's left? Good directional stability with peculiar handling and a flat floor—and that is precisely the difference between a Toronado and the more conventional automobiles it competes against.

Our test car was the Ultimate Weapon from the Oldsmobile camp. Every available option had been screwed, clamped, or bolted on at the Toronado works. Disc brakes, radial ply tires, Comfortron air conditioner, automatic speed control, stereo tape player, power bucket seats, electric windows all came along with the test car to warm us, cool us or confuse us. With them came a 400-hp Force-Air engine to keep everything in motion. And too, there was: tinted glass, head restraints, transistorized ignition, chrome wheels, tilt steering wheel, paint stripes and the list goes on. Come on, admit it. Our Toronado had more trick stuff than your car. It had to. The list kept going right up to $7,023.07 and 4655 pounds. Now are you convinced? You aren't? Maybe you'd like to hear about the sneaky little horn ring they've concealed inside the steering wheel rim. Wonderful device. All you have to do is squeeze it and it beeps. Never fails to go off when maneuvering into a tight parking spot or crawling in or out of the car. You could offend fewer people with rough, chapped hands or greyer than grey laundry.

When we heard about the optional 400-hp Force-Air engine we knew that nothing less would do: 455 cubic inches with dual exhausts, high performance camshaft, and fresh, outside air stuffed right up its air cleaner snorkle is a hard package to resist. We were right, it was great. Started and ran with never a temperamental moment and even broadcast a slightly irresponsible exhaust note just to assure the world of its presence. Still, regardless of engine size, neck-snapping acceleration can't be expected from a 4600-lb. car so we were satisfied with 15.7 second standing start quarter-mile times at 89.8 mph. Not inspiring but then it shouldn't alarm your insurance company either.

Provided you drive commuter fashion, the Toronado inspires driver confidence; but when you're hurrying through corners, you'd better know what you're doing.

The looks of the Toronado have changed significantly for the first time this year. It still doesn't look like anything but a Toronado, and while the profile remains the same it doesn't look like any Toronado we're used to seeing, either. Most of the stylists' efforts have been concentrated on the front end. They've cooked up a bumper-grille combination that appears even more unyielding than your local draft board and at the same time discontinued those incredible fold-up headlights which ranked above the 95th percentile on the ugly scale. Oldsmobile, never one to be outdone in the trick stuff department, has demonstrated clear cut engineering superiority over the rest of the industry with its latest headlight concealment apparatus. No more odd shaped trap doors in the sheetmetal or feebly rotating sections of the grille. That's for amateurs. By harnessing space-age science and several dozen amps from the battery, Oldsmobile has found a way to make the entire grille vanish at the touch of a button. And do you know what's behind the grille—other than the obvious four headlights? Would you believe another grille? No wonder Toronados cost a bundle.

Styling changes are not the only ones that have been made in the character of the Toronado since it was introduced in the fall of 1965. Instead of being a giant, 5-passenger sports car, the marketing philosophy has obviously been shifted a little to right of center so that now the car aligns itself more closely with its Thunderbird and Riviera competitors. The spring rates have been lowered and shock absorber control has been reduced so that ride quality is almost plushy when compared to the early models. Unfortunately, Oldsmobile got carried away with its shock absorber tuning with the result that the shocks do very little more than add weight to the car. Traveling freeways is like riding the high seas in a dinghy; you never stop floating up-and-down-and-up-and-down. A big improvement is the steering effort. No longer does the driver have any reason to believe that 62% of the weight is poised over the front wheels when he is parking. Even the body shake found in previous Toronados wasn't objectionable in our test car.

The engineers have been busy. Provided you drive in normal, commuter fashion the strongest impression of the Toronado is stability and predictability. As pointed out earlier, nose-heavy cars are like arrows, and even external disturbances like crosswinds have minimal effect. During standard, everyday operation, the Toronado inspires more driver confidence than any American luxury car we can remember.

Herein lies the problem. You recall what we said about overworked front tires and peculiar handling with fwd cars? All this becomes desperately plain when you start hurrying through corners, pumped up with confidence by the Toronado's impeccable low speed manners. You can count on the car to understeer but the degree of understeer varies widely with power application. Once set up in a corner, more power causes the path to straighten so that the car wants to fall off the road on the outside of the turn. Releasing the accelerator pedal does just the reverse; the Toronado dives toward the inside with alarming quickness. We think a giant, 400-hp, two-and-a-half ton Mini is a gas, but have some reservations about recommending it to citizens with cardiac conditions. A rule to remember—tire squeal from the overworked front tires is a vocal warning with an ample safety factor. If you always drive below the tire squeal point you'll be as safe as a cow in India.

While we are on the subject of tires, Oldsmobile offers radial ply balloons on the Toronado, primarily because of their long wearing characteristics. Our machine was wearing a monstrous set of 235R15 B.F.Goodrichs with fashionably skinny whitewalls. Of course there is no way to evaluate tire life on a road test but we did notice that the annoying traits normally accompanying radials were not at all objectionable. Ride harshness and low pitched, booming sounds are driving Detroit engineers nuts, but Oldsmobile apparently has a grip on the problem.

In the past we've voiced disapproval of the Toronado's brakes so we were sorry to see that no improvements have been made. Our first warning came during the acceleration tests when we faded the brakes completely while slowing for the first turn-off after four runs at Orange County Raceway. Smoke was pouring off the front brakes and we could just barely make the turn, normally no difficulty at all. All of this didn't establish a very promising precedent for the braking test. To obtain a quantitative measure of how bad the brakes really were, however, the test had to be made. Our pessimism manifested itself in the fact that we marked off 350 feet instead of the normal 250 for purposes of measurement. Drop back and let it happen. Much to everyone's surprise, the

TORONADO

Manufacturer: Oldsmobile Division
General Motors Corporation
Lansing, Michigan 48921

Number of dealers in U.S.: 3392

Vehicle type: front-engine, front-wheel-drive, 5-passenger coupe

Price as tested: $7,023.07
(Manufacturer's suggested retail price, including all options listed below, Federal excise tax, dealer preparation and delivery charges; does not include state and local taxes, license or freight charges)

Options on test car:
400-hp engine, power-assisted disc brakes, radial ply tires, chrome wheels, tilt and telescope steering column, accessory package, AM/FM stereo radio, tape player, power antenna, tinted glass, air conditioning, custom interior trim, deluxe seat belts, power windows, console, power door locks, power seat, recliner seat, head restraints, automatic speed control, high voltage ignition, bucket seats, cornering lamps, paint stripe, remote adjust outside mirror.

ENGINE
Type: water-cooled V-8, cast iron block and heads, 5 main bearings
Bore x stroke. 4.12 x 4.25 in, 104.8 x 106.9 mm
Displacement .... 455 cu in, 7450cc
Compression ratio .... 10.25 to one
Carburetion .... 1 x 4-bbl Rochester
Valve gear .... pushrod-operated overhead valves, hydraulic lifters.
Power (SAE) .... 400 bhp @ 4800 rpm
Torque (SAE) .... 500 lbs/ft @ 3200 rpm
Specific power output .... 0.88 bhp/cu in, 53.7 bhp/liter

DRIVE TRAIN
Transmission .... 3-speed automatic
Max. torque converter ratio .... 2.05 to one
Final drive ratio .... 3.07 to one

Gear	Ratio	Mph/1000 rpm	Max. test speed
I	2.48	10.9	54 mph (5000 rpm)
II	1.48	18.2	91 mph (5000 rpm)
III	1.00	27.0	120 mph (4450 rpm)

DIMENSIONS AND CAPACITIES
Wheelbase .... 119.0 in
Track .... F: 63.5 in, R: 63.0 in
Length .... 211.6 in
Width .... 76.4 in
Height .... 52.8 in
Ground clearance .... 4.9 in
Curb weight .... 4655 lbs
Test weight .... 4805 lbs
Weight distribution, F/R 62.2/37.8%
Lbs/bhp (test weight) 12.0
Battery capacity .... 12 volts, 75 amp/hr
Alternator capacity .... 660 watts
Fuel capacity .... 24 gal
Oil capacity .... 5 qts
Water capacity .... 19 qts

SUSPENSION
F: Ind., unequal length wishbones, torsion bars
R: Rigid axle, single leaf semi-elliptic springs, 2 shock absorbers per wheel

STEERING
Type .... Recirculating ball, power-assisted
Turns lock-to-lock 3.4
Turning circle .... 42.9 ft

BRAKES
F: .... 11.3-in power-assisted vented disc
R: 11.0 x 2.0-in power-assisted cast iron drums
Swept area .... 360 sq in

WHEELS AND TIRES
Wheel size and type .... 15 x 6-in, JK, chrome-plated, stamped steel, 5-bolt
Tire make, size and type .... BF Goodrich 235 R 15 radial ply, tubeless
Test inflation pressures ... F: 24 psi, R: 22 psi
Tire load rating ..... 1680 lbs per tire @ 24 psi

PERFORMANCE

Zero to	Seconds
30 mph	2.6
40 mph	3.8
50 mph	5.4
60 mph	7.7
70 mph	10.0
80 mph	12.7
90 mph	16.0
100 mph	21.0

Standing ¼-mile .... 15.7 sec @ 89.8 mph
80–0 mph panic stop .... 238 ft (.90 G)
Fuel mileage ..... 10–13 mpg on premium fuel
Cruising range .... 240–310 mi

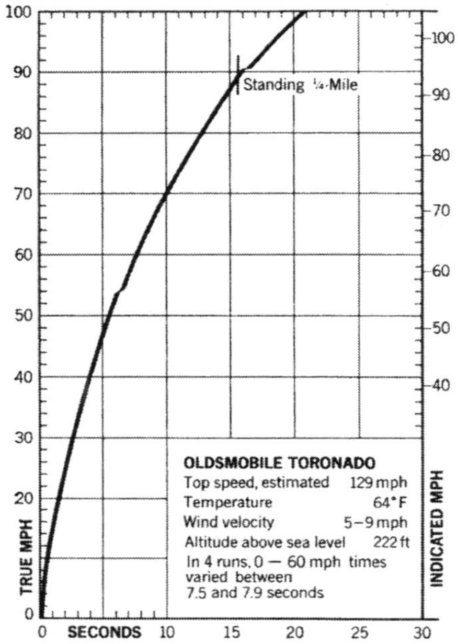

CHECK LIST

ENGINE
Starting .... Very Good
Response .... Very Good
Vibration .... Very Good
Noise .... Very Good

DRIVE TRAIN
Shift linkage .... Very Good
Shift smoothness .... Very Good
Drive train noise .... Very Good

STEERING
Effort .... Very Good
Response .... Very Good
Road feel .... Good
Kickback .... Very Good

SUSPENSION
Ride comfort .... Very Good
Roll resistance .... Very Good
Brake dive .... Very Good
Harshness control .... Good

HANDLING
Directional control .... Fair
Predictability .... Fair
Evasive maneuverability .... Good
Resistance to sidewinds .... Excellent

BRAKES
Pedal pressure .... Good
Response .... Good
Fade resistance .... Poor
Directional stability .... Poor

CONTROLS
Wheel position .... Excellent
Pedal position .... Very Good
Gearshift position .... Very Good
Relationship .... Very Good
Small controls .... Poor

INTERIOR
Ease of entry/exit .... Fair
Noise level (cruising) .... Good
Front seating comfort .... Very Good
Front leg room .... Good
Front head room .... Very Good
Front hip/shoulder room .... Very Good
Rear seating comfort .... Good
Rear leg room .... Good
Rear head room .... Good
Rear hip/shoulder room .... Good
Instrument comprehensiveness .... Poor
Instrument legibility .... Fair

VISION
Forward .... Very Good
Front quarter .... Very Good
Side .... Very Good
Rear quarter .... Fair
Rear .... Fair

WEATHER PROTECTION
Heater/defroster .... Very Good
Ventilation .... Poor
Air conditioner .... Excellent
Weather sealing .... Very Good

CONSTRUCTION QUALITY
Sheet metal .... Very Good
Paint .... Very Good
Chrome .... Fair
Upholstery .... Excellent
Padding .... Excellent
Hardware .... Excellent

GENERAL
Parking and signal lights .... Very Good
Wiper effectiveness .... Good
Service accessibility .... Fair
Trunk space .... Fair
Interior storage space .... Good
Bumper protection .... Excellent

OLDSMOBILE TORONADO

Lansing Landcruiser stopped painlessly from 80 mph in only 238 feet (.90G). Certainly an admirable stop for such a big car. Maybe all that smoke a little while ago was a figment of someone's imagination. With a little more optimism the second stop was approached. This time 270 feet at .79G. The fade was back and the driver was busy keeping the car in one lane. The third stop was the one. Describing the situation as mere violent instability would be too flattering. But for the quickness of the test driver, the guard rail along the strip was as good as gone, and a clean, one-owner Toronado would have soon been in need of cosmetic restoration. The power-assisted, disc-braked Oldsmobile Toronado finally came to rest after three breathtaking cycles of sideways oscillation with nothing worse than outrageously smelly brakes and flat-spotted tires. It was a relief. Admittedly, the first stop was exceptional for such a big car but the fade encountered during the following stops and the disastrous lack of directional stability are totally inadequate for travel on crowded, high speed throughways. Instead of piping cold air into the carburetor, Oldsmobile would do well to direct it toward the brakes.

Bewildering complexity is the design theme in the Toronado instrument panel. Right in the center, above the steering column, lives an incredibly simple slide rule type speedometer so small that it goes almost unnoticed in this cluttered field of irregularly shaped flashing devices. After a few moments of visually sorting things, we came to the awesome conclusion that, except for the fuel gauge, *the instrument panel contains no instruments.* It's really true. An impressive array of red, green and amber flashing lights are being brought to you through the courtesy of the Oldsmobile Division for your viewing pleasure. Our luxury car is a whole lot neater than yours. Even though these lights are frequently referred to as idiot lights, some are quite specific in their instructions. All during the test we drove in perpetual anxiety, wondering what would happen if we were a micro-second slow in complying with the "stop engine" light. A comprehensive selection of knobs—shaped like those you would find on your kitchen range—has also been provided for manually starting all kinds of wonderfully automated events in operation. The knobs want to be pulled, turned, or both, but there is no way to tell until you try. An example of the automation is the Comfortron air conditioner. A measure of Oldsmobile's pride in this contraption is its $500 price tag. Its claim to fame is simple and praiseworthy. Just set the desired passenger compartment temperature on the dial, move a lever over to one of the ON positions and your comfort is assured forevermore. If the world is hot, it air conditions. If the world is cold, it heats. If the world is just right, presumably it just nods in agreement and you never have to give it a tip.

In keeping with the little bit of everything in a luxury car concept you'd expect to find a half dozen pieces of wood incorporated into the interior decor, wouldn't you? Well, not real wood. Maybe imitation plastic boards or simulated kumquat or something, but the effort has to be made. Olds' stylists have dutifully glued their required quota of wood pieces on the Toronado. We have to admit that the state of the art in plastic wood is highly developed. Sometimes the only way we can expose the fakery is to tear up a corner of the part in question for closer scrutiny. It's just part of our comprehensive road test, however, and we feel compelled to report that Toronado wood does not grow on trees. We were able to disqualify the shift knob immediately because everybody knows kumquat trees do not have mold parting lines. Regardless of whether or not the wood is real, the Toronado reflects genuine quality. The panels fit well, the upholstery had only the designed-in wrinkles and everything was rattle-free. The only shabby areas on the whole car were the bumpers, particularly the front, where die marks were visible under the plating.

We are always amazed at the glorious but non-descriptive names the sales people attach to the extra cost options. The exterior of the test car was adorned with a GT Stripe which could be almost anything. In fact, it turned out to be a pair of parallel pin stripes which start on the roof, above the door, and wend their way along various parts of the body until they end up under the front bumper. We liked them, but we will never understand what makes them GT.

What is the Toronado, anyway? Even Oldsmobile will admit that it has never been a howling success in the sales reports. Why not? It's the only thing in its class without a "me too" approach. Except for the brakes, we could grow very fond of it. So why hasn't the car buying public taken it to their bosoms? Maybe it's because even Oldsmobile doesn't know what to do with it. Olds created it, pushed it out into the world and then stood back, wringing its corporate hands, waiting to see if it sank or swam. There have been times when Olds even pretended it wasn't there. They certainly make no attempt to capitalize on its great strength, its uniqueness. In searching through the Owner's Manual we found not one reference to fwd. That seemed curious. Front-wheel-drive has limitations but it also has advantages. Oldsmobile must have thought so or it wouldn't have built the thing in the first place.

Why should the world embrace the Toronado when even its mother has reservations? ●

TORONADO COMES on strong where going is slickest—combination of front-wheel drive, Goodrich 990 "super-radial" 9.15-15 tires and 60/40 front weight bias all contribute to remarkable traction and stability. Front tires look low, but actually are at specified 24 psi.

TODAY'S TORONADO

Expensive, luxurious, excellently engineered—and with virtues it alone has among American cars

FIRST OF ALL, let us firmly disabuse ourselves of the notion that a luxury car is no more than a gussied-up standard model, with a bit more foam in the seats, a somewhat quieter ride, and a few more mechanical amenities. It is not. A true luxury car has a unique identity, style and presence that lift it well above the common ruck and rubble whether on the boulevard or in the driveway. The Toronado is a luxury car and, to set it even further apart, it is a *sporting* luxury car.

Toward supplying a test car that would be all a Toronado could be, as well as a car to satisfy the most discriminating luxury/performance-oriented-owner, Oldsmobile gave us full run of the option catalog. When our spree ended, we had "spent" well over $7000—nearly half again the list price of the basic car.

CAR LIFE doesn't hand out Engineering Awards lightly, or on an automatic annual basis. They are given only to recognize broad and unusual achievement in the automotive engineering field, rather than the regular development of materials and components. Our most recent Engineering Excellence Award was given in November, 1965—to the Oldsmobile Toronado, for the successful application of the front-wheel-drive principle to a modern, high-performance automobile.

Since its introduction we have watched the evolution of the Toronado, and also noted that a good many U.S. designers have found its long hood, low lines and tucked-in rump attractive enough to emulate. Styling changes in the Toronado have not been extreme or, in our opinion, necessary: its configuration and design balance remain striking and thoroughly current. Front and rear aspects of the car have been altered somewhat, with the 1966-model headlights removed from their hood-pocket lairs and placed behind the hydraulically lifted grille screens, permitting the long, smooth hood lines to flow unbroken all the way from front bumper to windshield. The windshield wipers are hidden away in a special niche and the Toronado's rear has been divested of some unnecessary molding.

The major change for 1968, however, lies in a major displacement increase—from the original 425 cid to an even more potent 455, giving a maximum 375 bhp at 4600 rpm and 510 ft.-lb. of torque at 3000 rpm. This is now the standard Toronado engine, and drives through a low-ratio 3.08:1 differential. But our test car came equipped to deliver yet more power, with the optional (at $210) high-performance "force-air" induction system bringing maximum horsepower up to an even 400, in this case working through a 3.36:1 final drive. As the data panel indicates, performance was thoroughly stimulating—especially in view of our 5000 lb. test weight.

It is notable that the Toronado's 60/40 front-to-rear weight distribution, which would be considered excessively front-biased in a rear-drive car, was found optimum for good traction during its seven-year period of development. Weight transfer under hard acceleration lightens the front end, limiting traction of the driving wheels in a front-wheel drive configuration—an area in which rear-wheel drive enjoys a natural advantage. A strong front weight bias is necessary and beneficial to traction in the Toronado.

This condition, however, definitely is not beneficial to the front brakes, which must bear the brunt of tremendous loading under hard deceleration. Despite Olds engineers' early statements that disc brakes were not needed on the Toronado, they have been offered optionally this year, and we had them installed on the test car. A series of hard panic stops proved that front discs, at $80, are the best safety investment a Toronado buyer can make. Their stopping rate is good at 29 ft./sec.^{2}, without any loss of control or fade through eight stops. This is a remarkable improvement over the braking performance of the same car with drums all around (CL, February 1966).

Inside the Toronado, the first impression is of restrained opulence, the second of strength and purpose. The seats, individual buckets in the test car, are receptive but firm; the buttons, levers and switches are attractive to the eye but large and convenient enough for efficient use; the carpets are soft but not plushy; and the instruments themselves have a rather grand simplicity. The speedometer, in particular, is of the often-criticized "rolling-barrel" type, yet is big and well-damped enough for quick, accurate use.

The doors are extremely wide and sweep a bold, threatening arc in the parking lot, but do permit reasonably decorous rear seat entry and exit. Of course front-wheel drive does away with the necessity for a drive shaft tunnel down the middle of the car. With a front bench seat and steering-column control of the automatic transmission (only automatic is available in the Toronado) this would permit a flat floor area large enough for three adults to be readily comfortable. However, in the test car an optional (at $47) "transmission" console took up all the space between the bucket seats and ran well forward into one of the grandest front parlors in autodom.

The Turbo Hydra-Matic transmission is dependable, smooth and unobtrusive. The power flow mounts quietly up through its ratios with no whirring or hesitation between changes, a tribute to the engineers who labored for years over the design of this essential fwd component.

LOW, SLEEK silhouette of Toronado is result both of stylists' use of long horizontal lines in window, roof, and body area, and necessarily long hood containing both engine and automatic transmission. Hideaway opening for windshield wipers is visible just ahead of windshield.

INTERIOR is successful blend of masculinity, comfort and incorporation of safety elements, with plenty of tuck-and-roll vinyl, reclining right front seat and controls buried in door paneling and armrests. Inside of steering wheel rim is squeezed to sound horn.

1968 OLDSMOBILE

TORONADO 2-DOOR HARDTOP

DIMENSIONS

Wheelbase, in. 119.0
Track, f/r, in. 63.5/63.0
Overall length, in. 211.4
width 76.4
height 52.8
Front seat hip room, in. ......25.2 x 2
shoulder room 58.8
head room 37.9
pedal-seatback, max. 41.0
Rear seat hip room, in. 55.6
shoulder room 57.8
leg room 36.6
head room 37.3
Door opening width, in. 48.5
Trunk liftover height, in. 27.6

PRICES

List, FOB factory $4727
Equipped as tested $7023
Options included: Force-Air engine, Comfortron air conditioning, am/fm stereo radio/tape player, power disc brakes, wsw radial tires, C-D ignition, power windows and l.h. seat, chrome wheels.

CAPACITIES

No. of passengers 5
Luggage space, cu. ft. 14.5
Fuel tank, gal. 24
Crankcase, qt. 5
Transmission/dif., pt. 5.5/4.5
Radiator coolant, qt. 18.5

CHASSIS/SUSPENSION

Frame type: Unitized, front stub.
Front suspension type: Independent by s.l.a., torsion bar springs, telescopic shock absorbers.
ride rate at wheel, lb./in. 162
antiroll bar dia., in. 1.00
Rear suspension type: Fabricated solid axle on single leaf springs, two horizontal and two near-vertical telescopic shock absorbers.
ride rate at wheel, lb./in. 157
Steering system: Integral assist recirculating ball gear, parallelogram linkage ahead of front wheels.
overall ratio 17.8:1
turns, lock to lock 3.4
turning circle, ft. curb-curb 42.9
Curb weight, lb. 4660
Test weight 5000
Distribution (driver), % f/r 62.5/37.5

BRAKES

Type: Cast iron disc front, single leading shoe rear.
Front rotor, dia. x width, in. 11.3 x 1.9
Rear drum, dia. x width ....11.0 x 2.0
total swept area, sq. in. 360.0
Power assist: Integral vacuum.
line psi at 100 lb. pedal 1575

WHEELS/TIRES

Wheel rim size ....15 x 6JK
optional size ....none
bolt no./circle dia. in. 5/5.0
Tires: B.F. Goodrich Silvertown radial 990.
size ....235R (9.15)-15
normal inflation, psi f/r 24/24
Capacity @ p.s.i. ....6720 @ 24

ENGINE

Type, no. of cyl. ....ohv 90° V-8
Bore x stroke, in. ....4.126 x 4.250
Displacement, cu. in. 455
Compression ratio 10.25:1
Fuel required ....premium
Rated bhp @ rpm ....400 @ 4800
equivalent mph 118
Rated torque @ rpm .....500 @ 3200
equivalent mph 79
Carburetion: Rochester 1x4.
throttle dia., pri./sec. 1.38/2.25
Valve train: Hydraulic lifters, pushrods and overhead rocker arms.
cam timing
deg., int./exh. 24-81/74-33
duration, int./exh. 285/287
Exhaust system: Dual exhaust, reverse flow muffler.
pipe dia., exh./tail 2.25/2.0
Normal oil press. @ rpm. .40 @ 1500
Electrical supply, V./amp 12/55
Battery, plates/amp. hr. 90/75

DRIVE TRAIN

Transmission type: Three-speed automatic with torque converter.
Gear ratio 3rd (1.00:1) overall. 3.36:1
2nd (1.48:1) 4.98:1
1st (2.48:1) 8.33:1
1st x t.c. stall (2.05:1) 17.10:1
Shift lever location: Console.
Differential type: Front drive, spiral bevel gears in sprung differential.
axle ratio 3.36:1

TORONADO

The single most enjoyable accessory in the car was the am-fm/fm multiplex stereo radio. This fine instrument complements the quiet of the car with a clarity of sound that fully justifies its stiff $238 price. In the Toronado the stereo speakers are placed front and rear, rather than in the usual side locations, but in any case, the entertainment was rich and realistic.

The driving characteristics of the Toronado are surprisingly similar to those of rear drive cars, a point of pride to the engineers, who, of course, planned it exactly that way. It can rarely be detected that the car is pulling instead of pushing itself, except under the most severe cornering conditions. At the point of complete breakaway, the front end loosens up first and is difficult to steer back in, due to the extreme front weight bias. Of course the rear wheels cannot then be broken loose with the throttle to bring them around. In this situation, the best response is the instinctive one: get off the throttle, use the brakes, and drive the car back into line.

However, under all normal and most abnormal conditions the Toronado is an absolute jewel of a grand touring road machine. It wants very much to pull itself exactly where the front wheels are pointed—straight ahead down the highway or over the slipperiest bedevilments of rain, mud, snow or glare ice. In slick going the Toronado is the most maneuverable passenger car made in America.

It is not an ordinary automobile. When fitted out like the test car it is expensive, but has some strong attributes: high performance, dependable handling, strong masculine styling, the flair of front-wheel drive, and the availability of enough luxurious options to furnish a pleasure dome fit for a khan. Taken together they describe only the Toronado, an unusual and enjoyable automobile. ■

WELTER OF plumbing, wiring, ducts and pumps cloaks mighty 455-cid/400-bhp engine, which can propel 5000-lb. Toronado in sporting style while supplying energy for multitudinous accessories.

CAR LIFE ROAD TEST

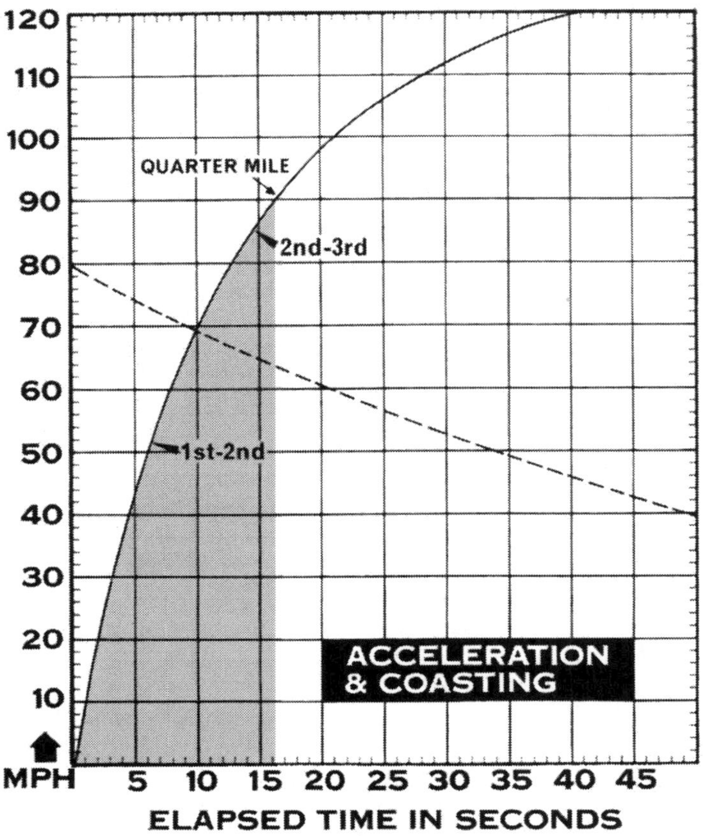

CALCULATED DATA

Lb./bhp (test weight)	12.5
Cu. ft./ton mile	128.8
Mph/1000 rpm (high gear)	24.6
Engine revs/miles (60 mph)	2440
Piston travel, ft./mile	1730
CAR LIFE wear index	42.2
Frontal area, sq. ft.	22.4
NHRA-AHRA class	F/SA-n.a

SPEEDOMETER ERROR

30 mph, actual	30.0
40 mph	38.6
50 mph	48.0
60 mph	57.9
70 mph	67.6
80 mph	77.9
90 mph	88.3

MAINTENANCE

Engine oil, miles/days	6000/120
oil filter, miles/days	6000/120
Chassis lubrication, miles	6000
Antismog servicing, type/miles	replace PCV valve/12,000, tune check/12,000
Air cleaner, miles	clean/12,000
Spark plugs: AC 44S.	
gap, (in.)	0.030
Basic timing, deg./rpm	10/850
max. cent. adv., deg./rpm	10/1800
max. vac. adv., deg./in. Hg	18.5/20.5
Ignition point gap, in.	0.016
cam dwell angle, deg.	30
arm tension, oz.	21
Tappet clearance, int./exh.	0/0
Fuel pressure at idle, psi	5.5
Radiator cap relief press., psi	15

PERFORMANCE

Top speed (5000), mph	123
Test shift points (rpm) @ mph	
2nd to 3rd (5200)	86
1st to 2nd (5200)	52

ACCELERATION

0-30 mph, sec.	3.5
0-40 mph	4.8
0-50 mph	6.3
0-60 mph	8.0
0-70 mph	10.0
0-80 mph	12.5
0-90 mph	16.2
0-100 mph	21.5
Standing ¼-mile, sec.	16.30
speed at end, mph	96.3
Passing, 30-70 mph, sec.	6.5

BRAKING

Max. deceleration rate from 80 mph ft./sec.²	29
No. of stops from 80 mph (60-sec. intervals) before 20% loss in deceleration rate	8-6% loss
Control loss?	None.
Overall brake performance	excellent

FUEL CONSUMPTION

Test conditions, mpg	8.7
Normal cond., mpg	8-12
Cruising range, miles	170-260

DRAG FACTOR

Total drag @ 60 mph, lb.	n.a.

Tenacious Toronados

(Above) Two-place version has low, lean look. Body sectioning in rear quarter has cut many inches off length.

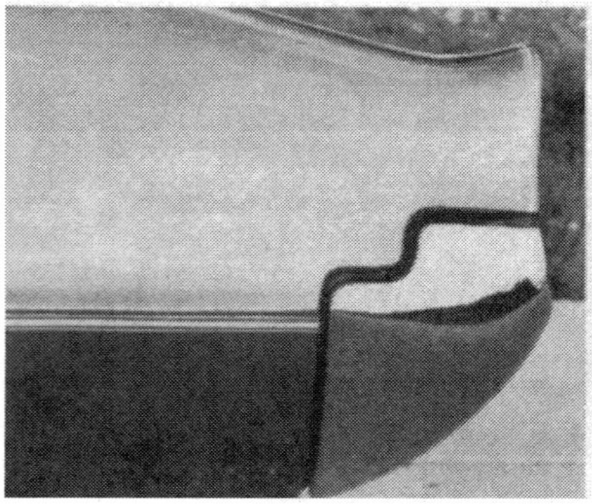

(Left) XX dash has switch to position brake and accelerator pedals in or out. (Below) Range of adjustment is evident.

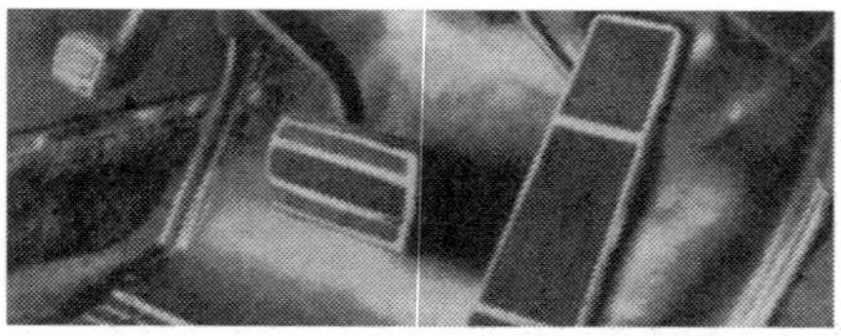

(Above) Both two-place and XX have individually contoured spoilers on rear deck. (Below) Shoulder harness fits into seat.

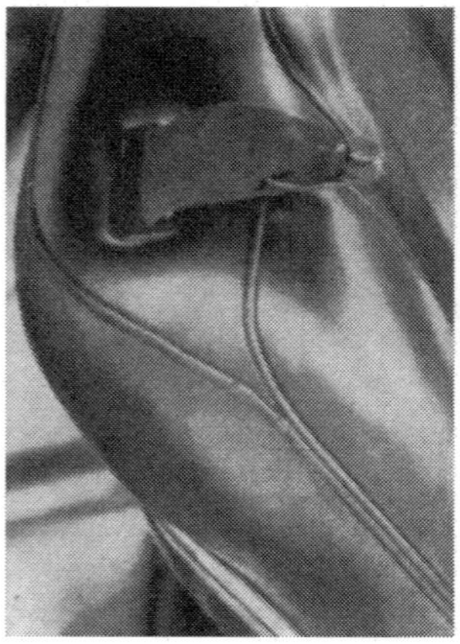

(Above) XX has special insignia to identify singular car. (Left) Model operates unusual roof console in two-seat Toronado. Power controls for individual seat and window adjustments are mounted in this overhead area between the two seats. A reading light and dome light are here, too.

Photos by George Foon, Oldsmobile Photographic

(Above) The hot XX exhibited fully controllable Toronado handling characteristics at Orange County Raceway. (Center) Another styling experiment features station wagon components on Toronado chassis, eliminating driveline bump. (Below) Three carburetors and special cam give XX plenty of "go."

Specifications Other Than Standard 1969 Toronado

Granturismo — 3.54:1 rear axle • Special camshaft • Special cylinder heads • Ram-Air • Special engine chrome • Toronado GT transmission — fast upshifts calibrated for quick and firm shifting • 110-inch wheelbase • 12.5:1 steering ratio • Stiffer suspension • **"XX"** — 3.54:1 rear axle • Three 2-barrel Rochester carburetors • Special engine chrome • Toronado GT transmission (same as Granturismo) • Special camshaft.

PERFORMANCE	Toronado "XX"
Acceleration	
0-30 mph	3.7 secs.
0-45 mph	6.2 secs.
0-60 mph	9.6 secs.
0-75 mph	13.9 secs.
Standing Start ¼-Mile	16.6 secs. 83.0 mph
Passing Speeds	
40-60 mph	5.2 secs. 380.6 ft.
50-70 mph	5.5 secs. 484.0 ft.
Speeds in Gears	
1st . . . mph @ rpm	48 @ 5000
2nd . . . mph @ rpm	80 @ 5000
3rd . . . mph @ rpm	102 @ 4500
MPH Per 1000 rpm	
(in top gear)	22.6 mph
Stopping Distances	
From 30 mph	24.8 ft.
From 60 mph	148.6 ft.

Styling precocity takes an upper hand at Oldsmobile

By Bill Sanders

"The times they are a changing," and nowhere is that song truer than at Oldsmobile styling. Your friendly, family Toronado may never be the same again. Olds stylists have been hard at work with their acetylene torches on the basic Toronado and . . . voila! We've got the "Granturismo," a two-place Toronado, the "XX," a sort of super Toronado and a Toronado station wagon.

The Granturismo is a designer's dream. Its wheelbase was shortened nine inches to a total 110 inches. Rear overhang was chopped five inches, making overall length 197 inches and height was reduced to a little over 50 inches. Door handles both outside and in are electrically operated. An interior roof console contains flush switches for power windows and seat adjustments. Stereo speakers are also overhead. The area left behind the seats has been used as a lockable storage compartment.

Toronado XX is a basically stock job. Stock in dimensions anyway. Under the hood the 455 cubic inch engine sports lots of chrome, a special high-duration cam and three two-barrel Rochester carbs. Inside it has adjustable accelerator and brake pedals that move to accommodate the tallest or shortest. Steering is extremely fast ratio. Shoulder harnesses have been incorporated into bucket seatbacks for convenience. The one and only XX is now being driven by Dick Smothers, but we had an opportunity to test it for a week before we delivered it to him. It was originally built as a show car, but still handles in the great Toronado style. Understeer was at a minimum with great power drifting capabilities. In the rain it was always under control. Although the big engine had lots of power options, including those three carburetors, performance was a little disappointing as you can see from the chart. Outside design was little altered except for a well designed deck spoiler.

A Toronado station wagon has been driving around in a lot of minds since the car was introduced. Grafting here and there with stock Olds station wagon components, the men with the torches have come up with an attractive, functional Toronado wagon. It's a great idea to utilize the unique capabilities of Toronado's front-wheel-drive.

Our friends from Oldsmobile have left us with one hang-up though. After whetting our appetites comes the crushing blow . . . the cars are one-of-a-kind. You can't buy 'em. The answer — letters to Oldsmobile, men! /MT

CITROËN VS. TORONADO

Fashion can be fun and how SCG test-matched the padded bra against a belt-in-the-back

A CITROEN?! YOU'RE GOING TO ROAD TEST A *CITROEN?!* And drive it for a *week?* Man . . . !"

"Yeah, why? What's the big deal, is it good or bad?"

"Oh, you'll see — you'll see!"

After getting that kind of response from 18 or 34 people, we were really hanging by our thumbs, a little apprehensive, but curious unto death. Since we (your road tester) had never driven a front-wheel-drive car before, we also borrowed a Toronado to run a little comparison test. Sort of a contest between French engineers and Detroit engineers. ("A Toronado?! You're going to road test a *Toronado?!* That's not a sports car!")

As it turned out, however, it wasn't much of a contest. Not that they aren't both damn fine cars — but they weren't playing the same game. It was more like the best of two different worlds from two different countries.

The Toronado was built with just one major, overwhelming goal in mind, a goal which influenced every component, system and shape in the entire package, and it shouldn't be surprising to realize that the corporation behind the Toronado exists for that one reason alone. General Motors isn't supposed to be artistic, or humane, or charitable, or loving; it is supposed to make a profit. So the Toronado was built to take the maximum number of dollars off a small number of people. (Stay cool, remember, most of those dollars go to wages for other people, who work hard for them.) Naturally Citroen wants to stay alive also, so they have a similar

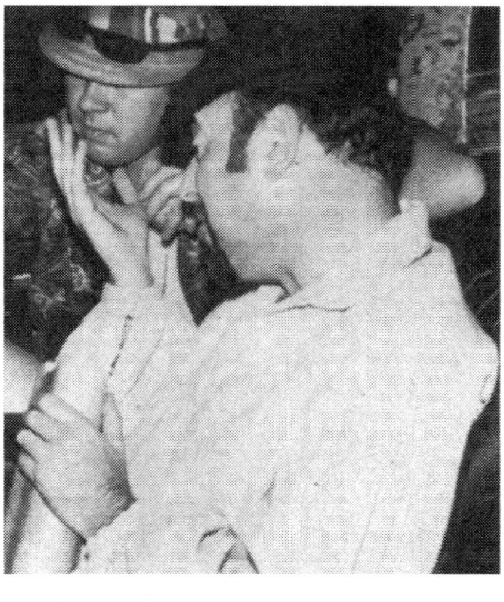

motive . . . but the method is a little different. They try to please a larger crowd and make a few less dollars per person. We aren't going to tell you which car is "better," because we don't have your outlook or your taste, but we are going to give the honest scoop on both so that they can be objectively classified into a "person's preference."

The Citroen is undoubtedly *the* unique sedan available today — in a word, a world's fair of engineering. If all the minds that went into this car had been working on something practical — like a race car — France wouldn't have to juggle numbers and letters like 3000 cc and CSI to win races. Most of the engineering features stem from the use of a central hydraulic system which furnishes power to operate all the flashy jiggers and makes the car as flexible as an acrobatic octopus. You know to which gimmick we refer — the suspension adjustment system that can raise or lower the front or back of the car over 6 inches either automatically, to compensate for load changes, or manually, to increase ground clearance (and ten times as often just to show off). The central hydraulic system also supplies juice for the power steering, power brakes, brake balance system, automatic clutch and air-oil suspension.

The Toronado, on the other hand, was made just as practically as possible from existing production lines, parts and past experience. The only really new systems (for GM) that required engineering were the chain drive to the offset, off-the-shelf transmission, and the constant-velocity front wheel U-joints. Outside of those requests to the engineering department, all the decisions apparently were made by the stylists and "human engineering" designers, who may be esthetic strumpets, but compared to Citroen, deserve a Nobel prize for contributions to creature and emotional comfort. The dictum on the wall at GM Styling is "I don't care if it's ugly, it sells!" and must be matched by one at Citroen Styling (a former one-man barbershop overlooking La Place Pigalle) that says "I don't care if it's ugly, it's aerodynamic," which *we* paraphrase to: "We don't care if it's aerodynamic, it's ugly."

Before we get a lot of mail from irate individualistic artists, let us elucidate. What we really object to is not just deviation from a norm, but total disregard for proportion and continuity. The front of the car is sleek and smooth — not unlike a four-eyed porpoise — but in full profile, the back of the body appears to be receding into the distance with every feature line taking its own course. Citroen proponents claim it is the lowest drag 4-door sedan in the world, which sounds like a big hedge erected to obfuscate Porsche. *Now* go at it, letter-writers, and tell us the Citroen DS-21 has a lower drag coefficient than any Porsche — and is, accordingly, prettier.

The difference in styling between Citroen and Toronado also illustrates a very practical problem: maneuvering in tight quarters. The Toronado isn't big, it's behemoth, 2 feet longer and 8 inches wider than the Citroen, which is a European luxury sedan. (However the Citroen has a 4 inch longer wheelbase — what did we say about proportion?) But you will find that the Toronado can be jockeyed almost as well because you can tell where the body ends, while the DS body slopes off and disappears from view long before it ends. Keep in mind that just because the front fenders clear, doesn't mean that the waist will, due to its low-drag-high-economy-wine-bottle body shape.

Also in the appearance, comfort and convenience department is our evaluation of the interiors. The seats in both cars are comfortable — in their own ways. The French way seems to be to engulf the driver in a foot of soft foam on all sides, which is somewhat like sitting on an air bag and leaves you out of touch with the vehicle except for what you can feel through the steering wheel. In comparison, the Toronado bench seems rather rigid, but it can be moved almost any direction except sideways with the power joystick, and the steering wheel likewise. Toronado has a power seat to make you comfortable, while Citroen has power suspension to do it.

Dashboards and controls are a shutout in favor of Toronado; the Citroen layout is a tragedy. A few years back a man named Dvorak made a time-and-motion study on typewriter keyboards and discovered that the arrangement was all wrong. So he designed an entirely different keyboard that was twice as fast, twice as efficient and easier to learn . . . and was never heard from again because no one was going to take time to change. Citroen wasn't paying attention. Their layout is more than just different, it's hodge-podge porridge. If they *must* scatter controls hither and yon, the least they could do is label them, and the most they could do is illuminate them so that you don't knock the shift lever into a lower gear while searching for the interior light switch to find out what gear you are (were) in. Further examples of their nonconformity are: manual shutoff only on the turn signal — pull it toward you; rotate a similar stalk one-half turn for park lights, one turn for headlights, pull toward you for dims; a choice of 12 scattered controls for heating and cooling; three adjacent unmarked switches for interior lights, warning lamp test and emergency flasher; the starter switch is a sector in the shift pattern and the mirror is in about the last place you would look. The speedometer is so cluttered with shift points, stopping distances in feet, and labels, that they only had room for 20 mph increments, and even a pilot with an "instrument" rating could better judge his speed by sticking his wet thumb out the window. The foregoing are hazardous distractions to anyone, but the brake pedal is probably safer where it is, once you get accustomed to it. *Immediately* to the left of the accelerator is an innocent-looking dimmer switch lying on the floor. Handle with care — it's really a tremendously powerful brake pedal in disguise. We are all left foot brakers because of the reduced reaction-time safety aspect, but after making a couple of stops by waving the left foot around feeling for the pedal, and finally jabbing the floor in desperation, we gave up and found the pedal very convenient to the right foot. Trying to keep our feet that close together while wearing tight pants was a big influence also.

The power steering system doesn't have much feel, because you probably don't want to feel what driven front wheels are doing anyway, but it makes up for that lack in its sound effects, when the hydraulic system goes "tisk-tisk-tisk" ("kiss-kiss-kiss" in French) with every tug on the wheel. Old-line Citroen drivers are familiar with all these things, and probably can't understand our consternation, but we'll bet that they leave the driver's manual open on the seat if they must trust the car to garage attendants. We can imagine some busy executive saying "Harrumph, I'm a busy man, son, I haven't got time to *think* while I'm driving." Our open letter to Citroen reads: "You may not like 'our' method of arranging controls, and it may not be the most efficient, but if you'd play ball you'd sell a lot more cars over here." But at least no one has ever had to lock a Citroen — who would steal one . . . or better, who in America *could* steal one.

The Toronado may have critical faults, but comfort isn't one of them, unless it lulls you into oblivion. It is actually a sports-pleasure machine, a stereo-hi-fidelity-FM-tape-deck classical sound room on wheels. When Stirling Moss and Juan Fangio were brought to the U.S. in 1966 to testify as expert witnesses in an automotive lawsuit, they were asked to test drive about 30 differ-

ent makes. All day long they hopped from car to car like children in toyland — and do you know what it was that awed these two international drivers? The sound systems available in American cars, and the musical variety on the radio.

Even if we *hated* Oldsmobile, we couldn't think of a nasty thing to say while listening to Vivaldi in stereo. And really, it's so luxurious that it's immoral. You sit down, turn it on, and it goes. But even if it didn't go, it would still be impressive because it has glamour, flash, zing, etc. The instrument panel is balanced over the steering column and looks like part of the set from "2001 — A Space Odyssey" except that there is a legible label on everything (we couldn't find the one that flashes "Life Systems Ceased" when the A/C malfunctions). There are a lot of controls, but at least they aren't in braille like in the Citroen. Speaking of a "lot of," brings up belts. Even with a carload of Naders there would be enough belts left over in the Toronado to buckle down Winsocki, and when they aren't being used they make the interior look like a trash can in a parachute shop.

It's soft living, though. Ever wonder what happened to the sports car nuts you don't see any more? They grew up, gave up, bought a pornographic towbarge like this and lived comfortably ever after. We put a lot of long, hard miles on the Toronado — we drove across Los Angeles twice *in one day* — and have just one last comment to make on the dashboard. Flock it. If the sun is shining anywhere in the state, all you can see in the windshield is shiny leatherette. It's dark leatherette, but it still reflects badly because of the low angle of the glass.

Top left, there are 19 keen features, 11 artistic faux pas, and 32 hidden errors in this picture. See if you can find them.

Top right, "Hello, my name is Hal. I am one of the world's most comfortable consoles. If you don't see anything you want, just ask me."

When it came to driver impressions of the two cars, again it was like two different worlds, with each car being excellent transportation for its particular planet. The Citromatic is a good trick-transmission for an economy car or backroad work such as rallying, but not quite smooth enough for a luxury sedan. Actually it is only an automatic *clutching* transmission, which requires you to select the proper gear, but eliminates the torque converter which digests the jerks and shocks (and gasoline) in a fully automatic box. The servo-mechanism that operates the clutch isn't bad — for a machine — but it took a week's work to figure out how to eliminate the jerks in the upshift (shift at low rpm where the engine doesn't have enough torque to jerk) and we never did learn how to start smoothly from rest. As a matter of fact, the slower the car goes, the worse everything functions. It's nice to have servos helping you do everything, but remember — they're *dumb;* they aren't easily adaptable to the wide range of conditions you get in normal driving. Trying to creep in traffic or parking, for example, is really a bundle of laughs when the clutch and brake servos get out of phase (their dynamic-response lags match frequency, but 180 degrees apart) and the car pulses along with the hiccups — assisted by the suspension servos trying to level the car on each jump. It doesn't happen frequently, but at critical times such as those ticklish moments when you suavely try to park in a tight slot at Bob's Big Boy.

The Olds Turbo-Hydra-Matic is something else. It may cost you a gallon of gas at every stop, but its shifts are as smooth and soft as a milk-fed frog's belly. For those who haven't driven a GM automatic since the years when you could anticipate a shift and make it sound like a "stick" by lifting your foot at the right time, forget it. This is the box they had trouble marketing because people thought it was broken — they couldn't hear it shift. Who needs a turbine engine when a trans can produce the same satin smoothness from a hammer-and-crank reciprocating engine? But don't forget — you pay for it, both in original cost and fuel gallonage. Perhaps we take economy a little too lightly in our road tests because our gas goes on the company credit card, and we might sing another song if mom said "you can't have it, Johnnie, unless you feed it and take care of it." Suddenly the Citroen has started looking much prettier when reflected in the unswerving eye of the gas pump counter window.

Once the DS 21 is in its own environment, such as at speed on autobahnen, Belgian blocks and California mud slides, it walks away from the Toronado, or slithers away as the case may be. On the same pot-holed back road where the Toronado practically came unwelded at 60 mph, the Citroen, with its amazing, adorable hydropneumatic

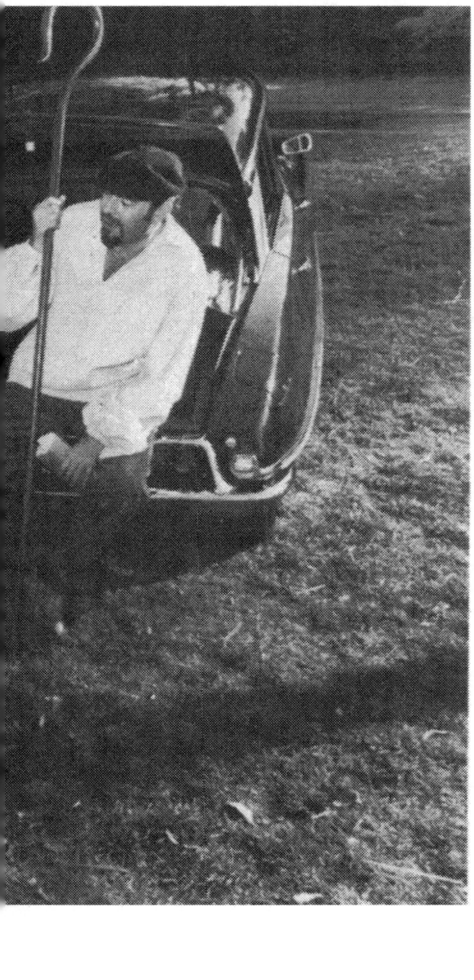

Left, "Voila, Zee Yankee's trunk is big enough for all his mink coats, credit cards, cameras, cigars, machine guns, anchor — while mine has just room for gold bullion."

suspension and inboard brakes (low unsprung weight) was comfortable and sure-footed at *80.* On one outing, when we had to detour around a rain-induced major earth displacement, we discovered that road conditions had no effect whatever on cruising speed. We slowed a little for the water fords to allow the car time to lift its skirts for wading — and out of fear that it might actually be a brim-full canyon instead of only a 3-foot-deep puddle. But dips, railroad crossings, chuckholes, sand, craters — no hang-up. The Citroen is highly touted as a good snow car also, which we tried to verify on a ski trip road test that was canceled by the highway patrol. The minions of the Law were dubious of even high ground clearance and front wheel drive pulling us through 22 feet of snow. No faith.

The Toronado on a freeway is all peaches, honey and sunshine — until you turn off that aphrodisiac sound system and fall into reality, where there is a little fly in the honey. The ride is fairly firm for such a huge machine, firm enough for you to know where you are, but also firm enough that a lot of road shocks and vibrations come through. Our test car had quite annoying harmonic sympathies that seemed to be an out-of-roundness in the tires, as though someone had flat-spotted them in a panic brake stop, which isn't easy to do as you will see. More likely, the tires never were round, unless Detroit (Akron, maybe? — Ed.) has whipped that problem on all their lines. It's kind of extravagant to put a very good tire on this car anyhow, since, as rumor has it, tire wear on the front is outrageous. And well it should be, with the weight on the front tires only slightly less than the *total* weight of the Citroen. However, odd as it may seem, the weight distribution isn't much worse than the big-block pony cars we have tested, at 59 percent on the front, while the Citroen has a ridiculous 64 percent front weight bias when empty. But we find that weight distribution alone plays only a small part in the cornering capabilities of a production car.

Out at our favorite playground, the Digitek Proving Grounds skid pad, we laid down a lot of rubber trying to discern a difference in the two cars. Since all cars are asymmetrical we run both directions around the circle, and when we averaged the runs we couldn't find a difference greater than the repeatability of our driver, although the Citroen had a greater variation in left versus right turns.

What *was* interesting — and a lot of race drivers have wondered about this — was the opportunity to test the effect of center of gravity height on cornering. The Citroen suspension has practically zero camber change, so when you drop it down on the bumpstops the only cornering gain you get is from the lowered c.g. Could you believe that this simple operation increased cornering power from the 0.66 g baseline to 0.71 g lateral acceleration. This method isn't recommended for racing, however, because if the bumps on an actual race course didn't destroy the gain they would probably destroy the car and your delicate duodenum.

While on the skid pad we also got some interesting notes on handling characteristics of front wheel drive in general, as the two cars were practically identical in amount of understeer and dynamic response. Foolproof. They don't seem to have a lot of steady-state understeer, but then it isn't needed for front-wheel drive because throttle application always increases understeer. And if you find yourself plowing off the verge, back off on the throttle — all the way if you like — the car will just tighten up the turn with no tail hang-out.

Outside of ten-tenths cornering, however, it's really difficult to tell which wheels are driving, unless you are on a very low coefficient surface. If the fact wasn't driven into your consciousness by advertising, 90 percent of the buyers of these cars would never know they had front-wheel drive. If you don't drive madly in the snow, who knows — who cares? Toronado prob-

Citroen DS-21

PRICE

As tested .... $5705 (POE West Coast)

Options ... Leather upholstery, electric defroster, air conditioning, tinted glass, air horn, AM-FM radio with roof antenna.

ENGINE

Type .... Four-cylinder, in-line, water-cooled

Head .... Aluminum alloy, hemispherical combustion chambers

Valves ... Ohv, pushrod/rocker actuated

Max. bhp @ rpm .. 109 @ 5500

Max. torque ...... 128 lbs. ft. @ 3500 rpm

Bore .... 3.54 in. (90 mm)

Stroke ...... 3.37 in. (85 mm)

Displacement ..... 133 cu. in. (2175 cc)

Compression ratio ... 8.8 to 1

Induction system ...... Single 2-bbl 28x36 DDE Weber

Exhaust system .. Dual, 4 into 2

Electrical system ..... 12 volt, alternator

DIFFERENTIAL

Type .... Spiral bevel

Ratio .... 4.38 to 1

Drive axles (type) ..... Double C.V. tripod universals

TRANSMISSION

Type .... Four-speed, full synchro, semi-automatic with hydraulic, governor-regulated clutch

Ratios: 1st 3.25
2nd 1.84
3rd 1.21
4th 0.85

STEERING

Type .... Rack and Pinion, power assisted

Turns (lock-to-lock) 3.25

Turn circle .... 36 ft.

BRAKES

Type ..... Inboard discs front, drum rear, variable ratio, dual system, power assisted

Disc diam: Front ...... 11.8 in.

Drum diam: Rear: ..... 10.0 in.

Swept area .... 387 sq. in.

CHASSIS

Frame ...... Unit construction

Body ..... Steel and fiberglass

Front suspension .... Independent, double leading-arm, hydro-pneumatic springs, integral shocks

Rear suspension ..... Independent, single trailing arm, hydro-pneumatic springs, integral shocks, self-leveling

Tire size ... 180-380/180 x 15

Tire type .. Michelin radial XAS

WEIGHTS and MEASURES

Wheelbase .... 123.0 in.

Front track .... 59.0 in.

Rear track .... 51.0 in.

Overall height .... 58.0 in.

Overall width .... 70.5 in.

Overall length .... 190.5 in.

Ground clearance ...... 2:5 in. to 8.5 in.

Curb weight .... 3250 lbs.

Test weight .... 3550 lbs.

Crankcase .... 4 qts.

Cooling system .... 9.5 qts.

Gas tank .... 17 gals.

PERFORMANCE RESULTS

ACCELERATION

0-30 .... 5.0 secs.

0-40 .... 7.9 secs.

0-50 .... 10.8 secs.

0-60 .... 14.5 secs.

0-70 .... 18.7 secs.

0-80 .... 25.5 secs.

0-90 .... 38.1 secs.

0-100 .... 50.0 secs.

Standing quarter-mile .... 20.3 secs. @ 73.7 mph

Top speed, mph 103

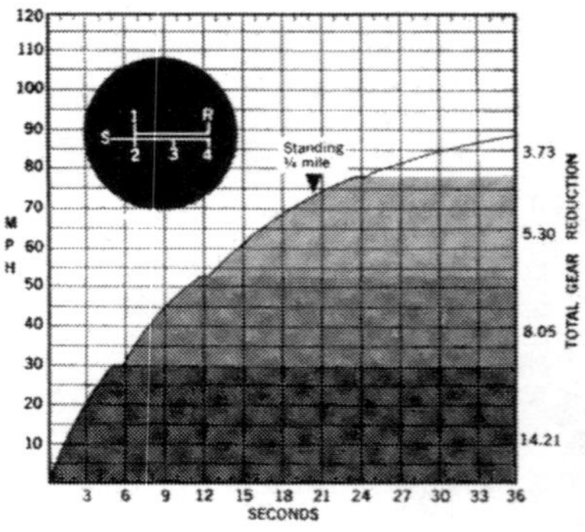

FUEL CONSUMPTION

Test .... 14 mpg

Average .... 18-20 mpg

BRAKING TEST

Deceleration, average .. 0.89 g

Fade .... None

SPEEDOMETER ERROR

Indicated	20	40	60	80	100
Actual	17	39	59	81	102

ably did it just to get rid of the transmission hump (sound of subdued applause) but that, quite obviously, was not Citroen's goal.

Another Citroenism that showed up in handling was its rather nebulous connection with the earth. The car pretty well takes care of itself, as we learned on rough roads, but you are never quite *sure.* That cerebral, but amorphous, supersoft suspension is always *doing* something. It's rather like riding on the back of an animal — say a hippopotamus — it gets you around all right, but you can never be positive just who is in control.

The straightaway performance tests can be summed up in one sentence. One goes but won't stop, and the other stops but won't go. You probably have a pretty good idea which car won the drag race; after all, it was *four hundred and fifty-five* cubic inches to 133, but the Citroen was further hampered by its lazy box. The brain has a control that can easily be adjusted to increase shift speed, but even at its fastest setting, the shift is a pretty phlegmatic automatic — and be careful, there is a neutral demon who lurks between each gear selection.

Now, where is that guy who said the Toronado is not a sports car? Remember what it did at Pikes Peak? We just have one question — how did they get it down? Our *official* brake test results say it took six stops to fade, at which time the smoke from the front brakes brought out a search-and-rescue squad from a nearby air base, but actually, by the end of each stop from 60 mph, the pedal could be pushed clean to the floor without locking the wheels. This *can't* be the correct solution to the anti-skid brake problem.

In contrast, the Citroen **can** be stopped from any speed all day long, as they perhaps have the best total brake system of any car in the world. The inboard front discs have integral air scoops from the nose to keep them from frying, the foot-operated emergency brake really *can* be used in emergency and not just for hill holding, the brake pads have an electrical contact to warn you when they are worn, a brake fluid level indicator is also in this circuit to a light on the dash, and bad pads can be changed in five minutes, but what is really distinctive is the brake balance adjuster. The ideal front-to-rear brake force distribution is not a static figure, but varies according to load and coefficient. By utilizing signals from the automatic suspension system, the brakes are automatically adjusted to rationalize braking under any load condition, a feature which is probably safer than the current rear-wheel-only anti-skid systems.

In summary, this comparison was a good example of stylists and designers versus engineers, and quite honestly, the results were not what we expected. The cultured artiste of our staff preferred the ugly old Citroen, and the technical personnel always wanted to drive the blandly engineered Toronado! But if we can't *predict* results, we are always more than happy to explain them once they arrive.

The best example of "overengineered" must be "Citroen." It seems to have been built *by* technical people, and intended *for* technical people, of which there are just so many in the world. And even at that, when tech people go home, perhaps they would rather forget their work, and be coddled instead, with glamour and style, as exemplified by the Toronado.

On the other hand, Citroen has always had a kind of reverse snob appeal. It's a thinking man's car — intellectual transportation. Not only do you have to think to drive it, but it's like wearing scholar's robes — it generates automatic respect. If we ever see a woman driving a DS and doing it well, we will probably come to attention and salute. The man who drives one seems to be saying, "I don't give a damn what you think, or how odd we look, I bought this car for its inner beauty and truth." Front-wheel-drive owners have a kind of camaraderie, they have clubs, swap stories about impossible journeys, and wave. But not Citroen owners. When they pass each other they just smile and discuss quantum physics by mental telepathy.

Oldsmobile Toronado

PRICE

As tested	$6486 (FOB Lansing)
Options	AM-FM radio, air conditioning, power seat, power brakes.

TRANSMISSION

Type	Turbo Hydra-Matic 400
Ratios: 1st	2.48
2nd	1.48
3rd	1.00

ENGINE

Type	V-8, water-cooled
Head	Cast iron
Valves	Ohv, pushrod/rocker actuated
Max. bhp @ rpm	400 @ 4800
Max. torque	510 lbs. ft. @ 3000 rpm
Bore	4.12 in. (105 mm)
Stroke	4.25 (108 mm)
Displacement	455 cu. in. (7356 cc)
Compression ratio	10.2 to 1
Induction system	Single 4-bbl. quadra-jet
Exhaust system	4 into 2
Electrical system	12 volt, alternator

STEERING

Type	Variable ratio, power-assisted
Turns (lock-to-lock)	3.4
Turn circle	43 ft.

BRAKES

Type	Drum, dual system, power-assisted
Drum diam: Front	11 in.
Drum diam: Rear	11 in.

CHASSIS

Frame	Perimeter, ladder type
Body	Steel
Front suspension	Unequal A-arms, torsion bars, tube shocks, anti-roll bar
Rear suspension	Beam axle, single leaf springs, tube shocks
Tire size	8.85 x 15
Tire type	Goodyear

DIFFERENTIAL

Type	Hypoid, sprung
Ratio	3.08 to 1
Drive axles (type)	Constant-velocity universals

WEIGHTS and MEASURES

Wheelbase	119.0 in.	Ground clearance	5.1 in.
Front track	63.5 in.	Curb weight	4960 lbs.
Rear track	63.0 in.	Test weight	5260 lbs.
Overall height	52.8 in.	Crankcase	5 qts.
Overall width	78.8 in.	Cooling system	18 qts.
Overall length	214.8 in.	Gas tank	24 gals.

PERFORMANCE RESULTS

ACCELERATION

0-30	3.0 secs.
0-40	4.7 secs.
0 50	6.6 secs.
0-60	8.8 secs.
0-70	11.5 secs.
0-80	14.9 secs.
0-90	19.2 secs.
0-100	25.3 secs.
Standing quarter-mile	16.4 secs. @ 87.0 mph
Top speed, mph	135 (est.)

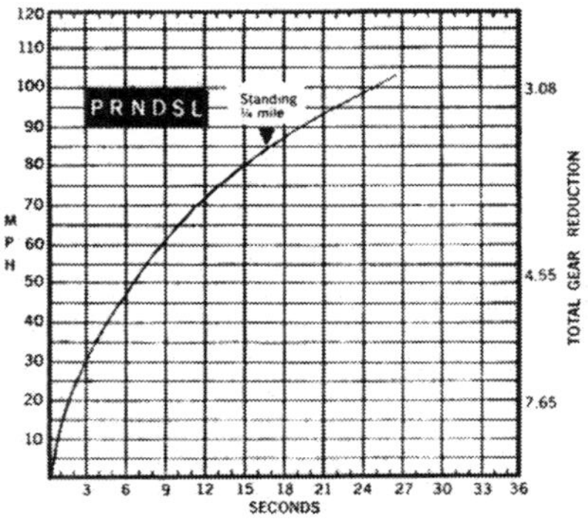

FUEL CONSUMPTION

Test	9 mpg
Average	12-13 mpg

BRAKING TEST

Deceleration, average	0.77 *g*
Fade	After 6th stop

SPEEDOMETER ERROR

Indicated	30	40	50	60	70	80	90	100
Actual	30	39	48	59	69	79	90	101

OLDSMOBILE TORONADO

• Oldsmobile, long the dark hope of all those who hope to escape big car tedium, has abdicated its chosen role. It has sold the Toronado down the river for reasons that are valid only on the tight grid of rules which govern inter-company competition in Detroit. It works like this: In the past, Rivieras, Thunderbirds and Lincoln Continentals have strongly outsold the Toronado and, with that information for its computers, Oldsmobile decided that the Toronado was wrong and the others were right. To rectify the situation, Oldsmobile has now fallen into step behind the competition with a Continental replica of its own. Naturally, it's greater in every dimension—it's longer, wider and taller—than past Toronados and stands as another monument commemorating that great series of American luxury cars which can be recognized by their small, glass-walled cabs mounted atop enormous, barge-like bottoms.

To the Toronado's credit, it has a clear aptitude for imitating the behavior as well as the appearance of a Lincoln Continental. It floats over humps and bumps with well near zero discomfort, always speaks with a soft voice and then only when spoken to. It is the very model of a modern transit capsule.

The question is, will anybody buy it? Presumably, there is a finite demand for this type of car and those who wanted one could have already found it in a Thunderbird or Continental or whatever. Now there is an extra mouth to share the same slice of total-market. And those who wanted a *real* Toronado have been abandoned.

It has always been our contention that the old Toronado was the unfortunate victim of corporate ambivalence and vacillation. The car simply wasn't Toronado enough. Remember back to the mid-Fifties—the days of the legendary Chrysler 300. *There* was a man's car, bold and strong, and in the public's eye it could do everything but leap tall buildings. The whole Chrysler line thrived on its reputation. The first Toronado was *almost* like that. It was as masculine as Michaelangelo's David, it felt tough and it had the fabled front-wheel-drive, a system universally recognized as Good. It was a unique car and America's eyes were upon it. But Oldsmobile backed out of the spotlight. Instead of enhancing the Toronado's attributes, it went the other way. The styling was mellowed, the ride softened and nobody talked about front drive. The Toronado character faded away. And customers' zeal faded with it.

We get the idea that Oldsmobile doesn't know quite what went wrong but the new Toronado is the final step in a kind of speedy but discreet about-face whereby Oldsmobile gets out of the freak car business and back into the luxury trade where it figures it belongs. And it is quite natural, in this kind of evolution, that the Toronado should share a greater number of parts with the rest of the Oldsmobile line. For 1971, it will be built on a full frame, rather than with a front stub as before, and the chassis is virtually identical to that of the 88. The leaf spring rear suspension is gone, replaced by the same 4-link arrangement used in the other big cars.

But considering the personality of the new Toronado, it is likely that the customers will be far more concerned with the interior. And those seeking luxury will know that they are in The Presence when they behold the broad gilded panels in the dash. It might be enough to keep their minds off the cruel lack of leg room in the rear seat. In fact, it might even be enough to make them forget that the Toronado is not a genuine Thunderbird or Lincoln Continental or Buick Riviera or whatever they've preferred up to now.

Dimensions	**1970**	**1971**
Wheelbase	119.0 in.	122.3 in.
Track, F/R	63.5/63.0 in.	63.5/63.6 in.
Length	214.3 in.	219.9 in.
Width	78.8 in.	79.8 in.
Engine		
Standard engine	455 cu. in., 375 hp. V-8	455 cu. in., 265 hp. V-8*
Compression ratio	10.25 to one	8.5 to one
Max. option	455 cu. in., 400 hp.	NONE
Compression ratio	10.25 to one	N.A.
Tires		
Standard	J78-15	J78-15

*SAE as installed

TORONADO

Toronado

Like its companion the Riviera, with which it shares the same body, the Toronado has changed very little in '69. A new grille and rear end bumper and tail light treatment set it apart from the '68 on the outside.

Interior treatment is the same as in '68 with front bench/bucket seats similar to those used in the Riviera. Three people can fit comfortably in the front seat although the Toronado is narrower than the Riviera. A folding center armrest aids driving comfort when only two are in front. Of course, the absence of a driveline hump on the front-wheel-drive Toronado gives a much more comfortable seating arrangement when six people are traveling.

The Toronado instrument panel is similar to that used in '68 with some of the switches and instruments moved around. Still a hallmark of the Toronado is the rolling drum speedometer, which isn't our favorite.

Cradling the biggest powerplant of all five cars, the Toronado still goes with the 455 cubic inch engine. The standard engine is rated at 375 hp. Our test car had Oldsmobile's high lift cam option (W-34) which includes dual exhausts. Horsepower rating is the highest at 400, as is the whopping 500 lbs.-ft. torque. The second heaviest car after the T-bird, the Toronado puts all the torque to good use for a powerfully smooth ride.

Ride and handling qualities with the front-wheel-drive Toronado are still superlative. Cornering, no matter how hard or rough is always flat and tight. The Toronado smooths and straightens out curves like they weren't there. There is no rolling or pitching experience even in the most vigorous driving situations. The Toronado is a great car where winter is severe. With front-wheel-drive, operation in snow, sleet and rain is superior to conventional drive cars.

Interior front leg room is more spacious than the Riviera, and the lack of a driveline hump adds to the roominess. Rear seat leg and head room are about the same. The Toronado has a huge trunk with lots of luggage space as the spare is placed so far forward it's almost in the back seat.

A Toronado is a Toronado and the '69 isn't much different. It's similar to the Riviera in comfort and convenience, but is superior in ride and handling characteristics with Toronado's front-wheel-drive.

(Top) Newest Toronado has completely redesigned grille and chopped rear end. Optional "pale grape" paint job attracted many compliments. (Above) Dash is similar to that of Riviera, with Oldsmobile instruments. "Rim blow" horn is still great feature. Tilting and telescoping steering wheel is also great on long trips to alleviate fatigue. Olds has exceptional stereo. (Right) Toronado trunk is quite large.

ALMOST A LIMOUSINE

Riviera-Thunderbird-Toronado, three personal luxury cars for the junior executive who almost has it made

BY JIM BROKAW

The man who buys one of the three current personal luxury standards – Toronado, Riviera, Thunderbird – is like the quarterback who has a first down on the goal line. Success is just a couple of smart moves away. But, if you blow the chance, you have to settle for a field goal and quarterbacks don't get credit for field goals. (George Blanda excepted.)

These cars are expensive. If you have to check your budget to see whether you can hand'e the payments, you can't afford one. They're totally impractical for many family pursuits, which indicates that there is at least one other means of transportation sharing the garage. Beach trips with wet bottoms and sandy feet are out. Hauling around a couple of sweaty, dirty, teeny-weeny football players is out. Interior space is comfortable for four adults, adequate for five, and tolerable for six on a quick trip from the office to an expense account lunch.

Power won't peel rubber but it will get you around town quick enough to grab a ticket if you try. Handling is surprisingly good on all three, but not good enough to get careless on strange roads. Economy really isn't there in the true sense of gas mileage.

If these creatures seem to have all of the disadvantages of the other classes of cars and few of the advantages, what do they have that keeps them selling? Prestige. In spite of advertising claims labeling the mini-limos as sports-luxury, the thing that brings in the buyer is good old fashioned status. For something under $7,000 the owner can announce to the world that he has

arrived, and will shortly move on to bigger things and out of the neighborhood.

Once upon a time, just because your car was luxurious, you didn't have to suffer poor handling. Cadillacs even raced at LeMans in the early 50's. But, in the last two decades, domestic luxury cars have grown fat and lazy, content in wallowing their way about corners as they tote up to a half-ton of options to keep their inhabitants oblivious to the fact that they're in a car rather than a living room.

But our test of this year's crop of instant envy machines came up with a surprise – handling that was unexpectedly good for luxury vehicles.

Our test cars were each ordered with the "whole shot" – big engine, all power, automatic transmissions – each representative of the way the cars are usually ordered by the big buck buyers in this class. The Thunderbird has the smallest engine, in terms of cubes, if you call a 429 V8 small, and the other two GM cars, the Riviera and Toronado, both had 455 cu. in. V8s, rated at 315 hp and 350 hp respectively. A higher horsepower version of the Riv, the Gran Sport, is available, but since there are no "hot" versions of the T-bird and Toronado, we did not order it. Gone is the mushy ride, and tire-scrubbing noises while you're turning, gone is the slow sway on the easy corner. Updated suspensions and the phenomenal recent success of Mercedes, have blessed the affluent with a firm but comfortable ride and the ability to hustle around a corner in a level attitude.

Both Toronado and Riviera have revised suspension systems. A wide span front lower control arm permits better dive control while the rear suspension on both units is completely redesigned. One of the reasons for a complete suspension revision is the switch to a full perimeter frame. The partial frame with a "floating" rear suspension has been abandoned. While both GM products have the full frame, the rear end treatment is unique to each model, primarily because the front-wheel-drive Toro has less weight to worry about in the back.

Toronado uses a fairly light coil spring with a trailing arm track bar. Transverse links retain lateral stability, but the rear axle actually pivots around the forward mount on the longitudinal track bar. This set-up very effectively eliminates the small bump of last year's dead axle/carriage-spring layout. Unfortunately, there is a slight penalty. While it handles the harsh bumps with disdain, the Toronado tends to loaf on the more moderately sloped rolling ones.

The Riviera does not have to contend with some rear end unsprung weight so the Buick engineers have taken a slightly different approach. The four-link layout is basically the same but the whole thing is beefier, including the springs, producing a different rebound effect than Toronado. The 1-inch anti-sway bar up front does a great deal to improve handling on the corners as well as dampening out any roll movement imparted to the rear end. Small, short amplitude bumps produce a slight harshness but the larger ones do not generate any unpleasant rebound. Under heavy cornering, the Riviera still tends to understeer and plow a bit, but much less than expected and not at all uncomfortable. Liberal use of body snubbers keeps the ride free of unwanted road noise.

Since both the Toronado and the Riviera are built by GM, albeit different divisions, the comparison of the handling characteristics is quite revealing. Like all the front-wheel-drive cars, the Toro has an inordinate amount of weight up front, tending to induce massive understeer, almost to the point that you would have to start turning at the corner grocery store to make it in the driveway. The anticipated trip to the outside of the turn never occurs, however, because the tendency to plow is overcome by the natural tendency of a powered wheel to proceed in the direction in which it is pointed.

Thunderbird also has four-coil suspension but the lateral and longitudinal restraints are handled a bit differently. The front end has drag-strut type bars to absorb high frequency vibrations, and high impact shock, which it does very nicely. The rear end has three trailing control arms, two below the axle and one above, and a lateral track bar. The net result is a slightly firmer ride than the Toro or the Riviera, but much less roll control than either. The T-bird takes a set position going into a corner which can be a bit disturbing if done at too high a rate of speed, but once it takes a tack, it holds what it has all the way through. The Bird requires a bit of attention going into a corner at high speed, but produces no surprises after the initial turn is passed. Small, rough bumps are felt, but not to an uncomfortable degree, and the larger ones are traversed unnoticed.

The most dominant impression when driving or riding in any of the three cars is the feeling of isolation you get – isolation from the road surface, isolation from any feeling of acceleration or braking. You could go from a concrete road, across a metal bridge grate, to a gravel road and there will be no vibrations coming to tell you that you're on a different road surface. While this will turn off "feel-of-the-road" buffs, this car-as-a-cocoon philosophy is great if you want to look upon your car as a place to unwind and relax, even while trying to drive 700 miles between sun-up and sun-down.

Riviera instrument panel lacks gauges, but clusters controls in easy reach. Right side dimensions are specifically tailored for a lady.

T-Bird dash is well instrumented and easy to read. Courtesy light under glare shield is excellent for map reading and other night games.

Toronado dash follows Riviera cockpit theme but lacks delicate balance and styling finesse of the dual curved approach favored by Buick.

Riviera's massive bumpers, bulging fenders and boat-tail rear deck constitute a return to a more substantial styling mode. Lateral distortion through rear window does not affect the driver's judgment of distance.

Thunderbird retains last year's styling with only miniscule chrome changes to enable the salesmen to distinguish the difference in the models. Durability of design is a new trend which may eliminate planned obsolescence.

Toronado bears a marked resemblance to the '70 Eldorado, but not by accident. Toronado fans will buy it and so will former Eldo customers who do not take to big brother's restyling for '71, so hope the marketing managers.

ALMOST A LIMOUSINE

Interestingly, Thunderbird is the only American car offered with radial ply tires as optional equipment, although our test car had the bias-belted wide ovals, which are more consistent with the cushiony ride concept. But, if you want luxury driving *plus* feel of the road, order radials.

In spite of the sporty styling and mammoth engines, none of the three cars will turn drag strip times that will scare any supercar owners. The Riviera was fastest, up to 60, turning an 8.8-second time compared to the T-bird's 9.2. Still, in one of those phenomenons known only to quarter-mile racing, the T-bird was fastest "through the eyes," turning a 16.25-second e.t. at 86 mph while the Riviera could only get it down to a 16.60-second e.t. at 88 mph. The Toronado was up at 16.90 seconds at 84 mph.

Speedometer error was most pronounced on the T-bird, reading 7 miles per hour faster than you were actually driving. The other two cars read a mile above or below your actual speed. The fact that brand new cars can have off-register speedometers should motivate you to get yours checked for accuracy before an officer does.

We entered into the braking tests with trepidation. While driving two tons of luxury car is great for comfort, there comes a time when you're going to have to stop in a hurry. It's then when you wish maybe your car was a little lighter.

All three cars stopped in a straight line from 30 mph, the T-bird stopping the quickest, and the Riviera second in braking ability. The story was the same from 60 mph, the T-bird taking 129 feet to stop, compared to the Riviera's 146 feet and the Toronado's 182. It was in the maximum effort stop from 60 that the heavy weight on the front end of all three cars, particularly the Toronado, demanded its penalty in controllability. All three cars began to swing sideways when the brakes were applied hard at 60 mph and, while the Riviera and T-bird took only one steering correction from the driver to prevent a spin, the Toronado took two fast lock-to-lock maneuvers to keep the back end from coming around, not at all in keeping with the machine's generally highly engineered nature.

The styling of the cars is the primary point of distinction. Each has a different theme, each imparts a unique impression and each generates a completely different attitude or feeling on the part of the driver. The Toronado is a refinement of last year's Eldorado. This is not an accident nor is it a money-saving ploy. The old Eldo was quite popular and people who buy Eldos tend to be quite firm in their opinions. Retaining a taste of Eldo in the Toronado opens the field to two types of customer. The ones who do not like the new design of the Eldo and did like the old, will be pleased with the Toronado. The Oldsmobile customers who couldn't afford an Eldo but liked them, will now be able to possess the styling they have been hankering after the past.

Riviera is a statement of what's happening. In spite of the fact that the car is a bold departure from what is current and accepted, it is by no means new. The rear end boat tail treatment bears more than a passing resemblance to a '63 Corvette, but its ancestry goes much farther back than that. Remember the Cadillac LeSabre in 1951? With the exception of the tail fins and the loop nose, the general body outlines are almost identical. This, reflecting perhaps a return to some of the solid values of earlier times.

Thunderbird doesn't have anything really new to offer other than last year's Bunkie Knudsen redesign. One gets the impression that T-bird is about to move off in a new direction but is now in a period of transition and hasn't yet made up its mind as to the intended path. Then too, it is part of the current durability-of-design concept being touted in Detroit; make it good and keep it around for awhile. Rumor has it that the current design for the Toronado will be with us for a good four or five years as well.

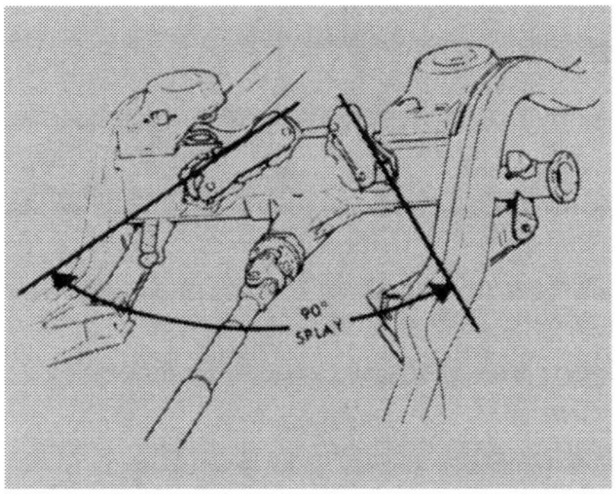

Riviera and Toronado share the same rear suspension design in different dimensions. Four link configuration permits greater roll and sway control with no sacrifice in comfort.

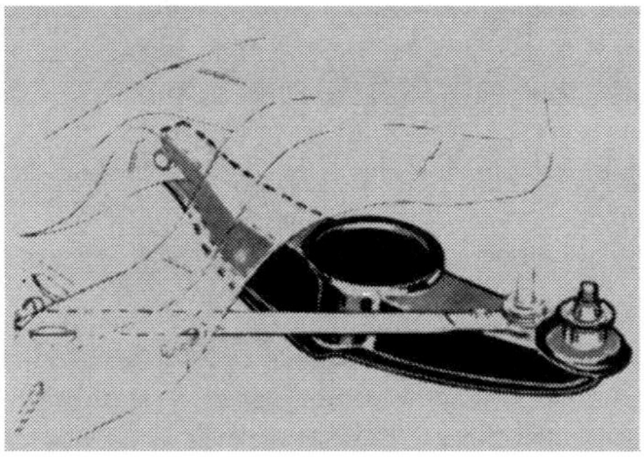

Above: Single mounted, narrow lower front control arm on Riviera has been replaced by heavier triangular piece (below) which reduces bump steer and increases stability.

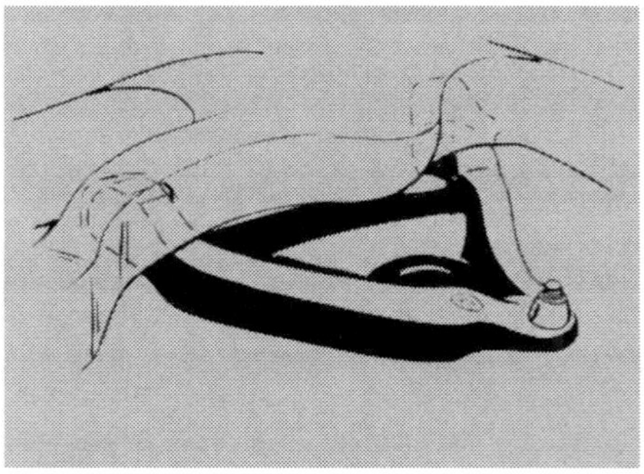

Interior styling presented more contrast between the three cars than their exterior. The Toronado was almost Spartan-looking, with the passenger's side of the instrument panel bare as a bone (for safety and to accept air bags should they be required) and all the controls hidden away and activated by mysterious electric motors. The Riviera came across with a bit more pomp and circumstance in the interior, with more plumped-up upholstery and a double-dip dash panel which molds itself around both driver and passenger. The Thunderbird's interior, though, was the Grand Palace of the lot. With its button-tufted brocade cloth upholstery, wrap-around rear seat and tunnel-like Cave of Love cockpit, created by the elimination of the rear quarter windows – it looked like something befitting Mae West or maybe even "Broadway" Joe Namath. With all its pizazz, the T-bird's sumptuousness couldn't match the practicality of Toronado's flat front floor – the unique advantage permitted by front wheel drive. One "problem" with the T-bird's interior was noted by a Swede who was invited to ride in the T-bird. He commented that it was typical of the American puritan ethic that "you have a car that looks like a bedroom but the seats don't fold down."

Overall, we feel that any of these cars fill the bill as far as being good combinations of comfort conveniences, handling quality and performance. You couldn't be any more comfortable in a Lincoln or Cadillac unless someone else was driving. Their air conditioners wouldn't make you any cooler nor would their seats be any softer. But the bigger-is-better philosophy still rules at GM and Ford styling and lots of people will still go on thinking that a car the size of a Riviera, Toronado or T-bird couldn't be as luxurious as a car three feet longer. We would even like to see the luxury approach tried on something three feet shorter than our test cars.

While all the attention paid to the Vega, Pinto, et al would seem to give the impression that the auto-makers have abandoned their Queen Mary-sized cars and taken to the life-boat-sized compacts, there are still thousands of consumers to whom "bigger is better" is a viable concept. You may call them the "establishment" derogatorily but, for establishment cars, the three we tested were cars worth owning, and even saving for. Tradition is a good thing-sometimes. /MT

MOTOR TREND SPECIFICATION DATA

SPECIFICATIONS	RIVIERA	THUNDERBIRD	TORONADO
Engine:	90° OHV V8	90° OHV V8	90° OHV V8
Bore & Stroke — ins.	4.3125 x 3.90	4.362 x 3.59	4.125 x 4.250
Displacement — cu. in.	455	429	455
HP @ RPM	315 @ 4400	360 @ 4600	275 @ 4200
Torque: lbs.-ft. @ rpm	450 @ 2800	480 @ 2800	375 @ 2800
Compression Ratio	8.5:1	10.5:1	8.5:1
Carburetion	4V	4V	4V
Transmission	3-spd auto, turbo hydramatic	3-spd auto	3-spd auto
Final Drive Ratio	2.93:1	3.00:1	3.07:1
Steering Type	Power, variable ratio	Power	Power
Steering Ratio	16.5-14.56:1	21.9:1	17.9:1
Turning Diameter (Curb-to-curb-ft.)	43.3	42.7	44.3
Wheel Turns (lock-to-lock)	2.94	4.0	3.4
Tire Size	H78-15 bias-belted	H78-15 bias-belted	J78-15 bias belted
Brakes	Power disc/drum	Power disc/drum	Power disc/drum
Front Suspension	Coil, shock, upper & lower A frame control arms	Coil, shock, upper & lower control arms	Torsion bar link stabilizer
Rear Suspension	Coil, shock, 4-link	Coil, shock, 3 control arms & lateral track bar	Coil, shock, 4-link
Body/Frame Construction	Body on full Perimeter frame	Body on full perimeter frame	Body on full perimeter frame

PERFORMANCE	RIVIERA	THUNDERBIRD	TORONADO
Acceleration 0-30 mph	3.3	3.9	4.4
0.45 mph	5.5	6.4	7.2
0-60 mph	8.4	9.2	10.7
0-75 mph	12.6	13.3	15.5
Standing Start ¼-mile mph	83	85.6	84.0
Elapsed time	16.9	16.4	16.9
Passing speeds 40-60 mph	5.1	4.1	6.4
50-70 mph	5.5	5.0	6.5
Speed in gears* 1st . . . . mph @ rpm	45 @ 4000	43 @ 4000	48 @ 4500
2nd . . . . mph @ rpm	77 @ 4000	74 @ 4000	80 @ 4500
3rd . . . . mph @ rpm	96 @ 3500	91 @ 3500	89 @ 3500
Mph per 1000 rpm (in top gear)	27.4	26	25.4
Stopping distances From 30 mph	29.8	27	30.0
From 60 mph	135.2	145	175.0
Speedometer error Electric speedometer	30 45 50 60 70 80	30 45 50 60 70 80	30 45 50 60 70 80
Car speedometer	31 46 51 62 72 82	37 52 57 67 78 89	28.5 43 48.5 59 69 79

*Speeds in gears are maximum speeds. (limited by the length of track)

1971 RIVIERA

Mfg. suggested retail price	$5,251.00
Max trac	91.57
Radio AM/FM	238.92
White walls	40.00
Climate control	515.73
Seat belts	26.32
Cornering lights	36.84
Speed alert	17.90
Fingertip windshield wipers	21.05
Tinted glass	49.47
Rear window defroster	31.58
Power seats (6-way)	105.25
Power windows	126.30
Cruise master	68.42
Electric door locks	47.37
Power brakes disc/drum	Std.
Power steering	Std.
Invoice	$6,667.72

1971 THUNDERBIRD

Two-door Landau base	$5,357.00
429-4V	Std.
Select shift Cruis-o-matic	Std.
Power steering	Std.
Power brakes disc/drum	Std.
Vinyl roof	Std.
Sequential turn signals	Std.
Remote mirror	Std.
Electric clock	Std.
Front cornering lights	Std.
Auto parking brake release	Std.
NOX emission control	Std.
Brougham cloth & vinyl trim	170.13
WSW tires H78x15	31.53
Convenience Check Group	
Courtesy lights & extra warning lights	
Seat back release — vacuum door locks	106.07
Tilt steering wheel	54.67
6-way power seat	103.89
A/C select-air	448.36
AM/FM stereo radio	157.47
Tinted glass	50.46
Power windows	115.46
Deluxe wheel covers	54.67
Invoice	$6,649.71

1971 TORONADO

Base price	$5,459.00
Corner lamps	36.86
A/C 4 season	437.08
Tinted glass	49.50
Belted WSW tires	36.86
AM radio	87.42
Rear speaker	18.96
Power trunk release	14.78
Convenience group	
Courtesy lights & vanity mirror	
Extra warning lights	21.80
Chrome door guards	6.32
Power windows	126.38
Power door locks	47.39
6-way power seat	105.32
Low fuel warning light	9.48
Automatic transmission	Std.
Power steering	Std.
Power brakes disc/drum	Std.
Dual Exhaust	Std.
Heavy duty battery	Std.
Clock	Std.
Rear cigarette lighters	Std.
Front center arm rest	Std.
Remote mirror	Std.
Invoice	$6,457.15

Olds Toronado Grand Turisimo

The Oldsmobile Toronado Granturismo is a specially crafted personal version of America's premiere front wheel drive automobile. It has been given a new, intimate proportion for two passengers. Powered by a high-performance version of the Toronado Rocket V8, it is even more agile than its production counterpart.

The Toronado Granturismo is a rolling display case for a host of design innovations. Under the skin, its wheelbase was shortened 9 inches, to achieve a GT dimension of 110 inches. Rear overhang was reduced 5 inches, bringing the overall length to 197 inches. The overall height was brought down to 50.4 inches. New rear springs were fabricated to meet the changed load requirements.

Under the hood, a high-performance 475.32-cubic inch engine of 4.219 x 4.250-inch bore and stroke, running a high lift cam and 10.25:1 compression ratio, provides brilliant performance. Carburetion is accomplished through three twin-throat Rochester units.

The exterior detailing of the Toronado Granturismo has been carefully executed to enhance the new proportion. A fine, hexagonal honeycomb texture is used in the grille. Slim windshield pillars have been specially formed to blend with the crisp roof line. A new belt line on the door and a new quarter window with transparent seal combine to create a European-type side window treatment. Custom door handles, of the style found on foreign coachwork, are flush with the body surface.

SLEEK STYLING – This expansive and sleek styling treatment of the Toronado side window gave the car a special look and set it apart from the ordinary Toronado of that or any other year.

At the rear a raised airflow deflector, which aerodynamically loads the rear wheels at speed, terminates the body form. Immediately beneath it are body ventilation outlets. Running lights define the car width at the sides. The exterior finish is selectively graded to accentuate the light and shade pattern created by the body forms.

Inside, a total philosophy of protective convenience has been followed, achieving a design treatment that is new, inviting, and comfortable. The interior, with the exception of the seat, is completely trimmed with molded panels. No metal garnish moldings are used, and all necessary hardware is recessed behind protective padding.

The seat is a special high-backed version of the Toronado's Astro-bench seat. Inertia reel shoulder harnesses combine total body restraint with comfort and ease of use.

Behind the seats is a lockable storage case to conceal valuable personal gear, such as cameras, shotguns, etc.

The instrumental panel cluster features an electronic, digital readout speedometer display which shows the driver his exact speed in large numerals. Warning lamps have been regrouped and increased to include: seat belt, low fuel, door unlocked, emergency flasher.

Above, a roof console incorporates flush switches for power windows and the 6-way power seat. Also there are a reading light and dome light.

The steering wheel is a special H-spoke, padded design adjustable for tilt and telescope upon release of a new, single-action lever on the column.

The door trim contains courtesy lamps to illuminate the exit area and sequencing flashers to warn approaching cars that the door is open. The door handle is recessed, and within it is an emergency manual release for the electric door locking system. Flush push pads are provided for the door release and door lock functions.

Interior ventilation is accomplished by exhausting the air through grilles at the rear edge of the window opening and carrying it rearward to the external outlets.

The Toronado Granturismo is equipped with a stereo radio and tape system and features four unique overhead speakers which place sound to the right and left of **both** the driver's and passenger's heads.

Design innovations are no accident. They occur when designers recognize a need and attempt to satisfy it. Cars such as the Toronado Granturismo are the medium through which these innovations are experienced, throughly tested, and evaluated before being adopted for production.

o

OVERHEAD CONTROLS — This was a styling feature that didn't find its way into the production Toronado's — overhead controls, which made the car seem more like an airplane cockpit than a family sedan.

SHORTENED BODY — The rear fender and rear deck were modified making the car shorter and giving it a "spoiler" lip and an overall short coupled appearance.

Toronado General Concept

by J.B. Beltz

EDITOR'S NOTE: At the time this was written, J.B. Beltz was the Chief Engineer of the Oldsmobile Division. Since that time he has become the current general manager of Oldsmobile. Like many general managers before him, Beltz got the top job at Oldsmobile after a long climb up through the ranks as an outstanding engineer. We are proud to reproduce this paper which was prepared for the Society of Automotive Engineers, and presented at that body's Automotive Engineering Congress, Detroit, Michigan, January 10-14, 1966.

This paper is a preamble to four others written by members of the Oldsmobile Engineering organization on the major phases of the Toronado design and development. The purpose is to provide general orientation in the arrangement and features of this new front wheel drive automobile, and to give insight into why and how the car was developed.

The Toronado venture was born of the desire to create a better automobile, one with more usable room and improved roadworthiness. The program was initiated because of our conviction that a continuing policy of offering something better in automotive transportation is essential to Oldsmobile's increased success. We were after a big step forward. This meant approaching the design unfettered by commitment to traditional arrangement.

The use of front wheel drive to permit combining all power and driving components into a single unit ahead of the passenger compartment, proved to be an important element in attaining that improved vehicle. As a result, the Toronado represents a new breed of contemporary automobile, an American Grand Touring Car.

The emphasis here should be noted. The important thing is the car as a whole, as a more enjoyable and efficient transportation machine. Front wheel drive was a way to get it.

Final realization of this objective in the Toronado represents thousands of creative manhours, one and a half million test miles and over seven years of work. The car culminates a combined effort by engineers not only at Oldsmobile but at General Motors Staff Activities and Allied Divisions in the areas of design, styling, testing, development, and manufacturing.

Outstanding contributions by Allied Divisions were made by Saginaw Steering Gear Division on the drive shafts and joints, and by Hydra-Matic Division on the drive chain system. As for General Motors Engineering Staff, they made important contributions in many areas of the development.

The original exploratory work on front wheel drive was done with an F-85 size car in mind. The first experimental car of this type was built early in 1960. The traction and handling characteristics were highly encouraging and it had good directional stability.

However, market studies had indicated that people in the so-called "Luxury Sports Car" segment of the market were especially interested in new things and were getting somewhat restive about being offered new cars which were new only in appearance. As a result, it was then decided that effort on front wheel drive would be concentrated on a full size high performance car.

From a design point of view, front wheel drive is a very clean way to lay out an automobile. When all the machinery is grouped together in the front, it permits a lot more freedom in designing space for the passengers. It also gives the chance to reduce exterior size. The car has a 4" shorter wheelbase, 6" less overall length and 1.5" less width than our regular rear drive Starfire hardtop coupe (Table I). This reduction in exterior size is aided by the more efficient use of inside space as the result of eliminating the floor tunnels and kick-up over the rear axle.

There are a great many possible ways to arrange the components in a front drive car. The important thing about the composite drive system on the Toronado is that it uses space most efficiently. This single unit, consisting of engine, transmission and differential occupies very little more space than the engine alone in a rear drive car.

Review of the major components and their arrangement in the car discloses many areas of achievement in design.

Beginning with the chassis, the frame of the car is unusual in that it ends behind the forward mounting of the rear spring. This arrangement permits isolation of the drive and suspension components from the body with rubber body mounts on the separate frame, while at the same time using frame integral construction at the rear of the car where space is at a premium.

The area of vehicle structure called for careful study. Relocation and drastic revision of the weight of the major components of the drive system resulted in a car with structural behavior entirely different from what we were used to. Resonant shake was a characteristic of our first pretest cars.

Vehicle structure occupied much of our attention through the entire span of the development program as we gradually learned how to best tune the new system. Modifications were required in front and rear suspension, frame, body structure and power plant mounting to finally get what we wanted. During this process the frame grew from a stub bolted to the front of dash to a complete structure running back to the rear spring front eye.

Single leaf springs and a stamped axle assembly are used in the rear suspension. The suspension is mounted through rubber at the spring eyes. Four rear shock absorbers are used, two vertical, two horizontal. The benefit of horizontal shocks is that they permit valving for best control of both vertical ride motions and fore and aft axle motions without the compromise usually required with single shock absorbers. The added horizontal shocks give good control of wheel hop when braking hard and make a worthwile contribution to the car's ride balance.

Front suspension is a torsion bar design. The torsion bars work from the lower arms and are anchored in a cross bar which is isolated from the frame through rubber. One set of torsion bars covers all car weights by the provision of carrying height adjustment at the anchor end, permitting accurate setting of carrying height in production and service. This eliminates the need for an inventory of different spring loads to handle various car weights.

A dual outlet exhaust system is used with the muffler located crosswise, ahead of the fuel tank and behind the rear axle. The flat, horizontal fuel tank with rear fill is located beneath the trunk floor.

Overall steering ratio is 17.8:1. This compares to 21.7:1 for our present production full size car. The steering column assembly uses an added universal joint to provide proper column angle.

Drive shafts on each side of the car transmit torque to the front wheels, through two constant velocity universal joints on each shaft. The right hand shaft assembly includes a rubber damper which cushions peak forces for maximum drive smoothness. This consists of a torque-in-groove arrangement surrounded by a rubber biscuit. The shaft stops metal to metal after 7 degrees of wind-up. It provides resiliency in the drive system for good transmission smoothness and desensitizes the system to lash effects.

The inboard joint takes in and out travel as well as rotational and angular movements. This is accomplished by an additional ball race which permits the entire joint assembly to move laterally as the suspension swings.

During the early development of the car, a number of brake systems were evaluated, including inboard and outboard mounted disc designs, with the objective of attaining the best brakes for the car's requirements. The best overall combination developed was an improved drum type. Cast finned brake drums are used which are cooled through openings in the wheel spider.

The cross section of the front and rear brakes show the relationship between wheel drum and spindles. The wheel spider and drum are offset out into the air stream. Wheel openings are designed for adequate air flow over the drum fins. This design has resulted in reduced brake temperatures, giving good fade characteristics and stability.

A great many different versions of the wheel were tested in the process of developing one with best durability and most effective openings for brake cooling.

Since driving torque and steering are both transmitted through the front wheels, care was taken in the selection of a tire that well suited the car's needs. Particular attention was paid to tire structure in tuning for good handling and tread pattern for excellent traction. The car uses an 8.85 x 15 "T-F-D" tire. The "T-F-D" designates "Toronado-Front-Drive".

Tire life with our regular 6,000 mile rotation is slightly better than attained with our conventional rear drive cars. It is not necessary to rotate the spare.

Located on the L.H. side of the car, the differential feeds torque to the left drive shaft directly, and to the right drive shaft through a cross shaft.

In order to meet space requirements and reduce internal friction, the design uses a planetary gear set instead of the usual bevel gears to produce differential action.

The transmission is new, adapting the Turbo Hydra-matic design to front drive. It is mounted lengthwise of the car on the left side of the engine crankcase. The differential is bolted directly to the transmission case. This arrangement keeps the transmission from infringing on the passenger compartment, and gives a flat toe pan and floor.

Our basic 425 cu. in. Rocket Engine has been modified to suit the front wheel drive car. Intake manifold and air cleaner are lowered to provide hood clearance. Exhaust manifolds are revised for frame clearance. Larger intake valves and a special camshaft provide additional horsepower. The engine is mounted 1.8" right of car centerline for adequate clearance to suspension components.

Mounted on the engine crankshaft, the converter feeds power to the transmission through a 2" wide chain drive. The chain has 3/8" pitch and a developed length of 46.5". It is positively lubricated by spit holes from the transmission lube system.

A durable and quiet chain drive between the engine and transmission was essential to the design of the car. A number of approaches were evaluated during the early stages of the development, including several gear drive arrangements. The chain was finally chosen as durability was excellent and we were never able to get satisfactory quietness with gears.

Conventional body construction is used except that the frame ends at the rear spring front eye so that the structure from there rearward is integral in the body. The rear spring rear eye is mounted to this structure through a rubber isolated shackle.

Unusually good entrance and exit conditions were achieved by using an extra long door and positioning it rearward in relation to the seat. This moved the lock pillar at the rear edge of the door farther away from the front seat, giving a wide area of entry to the rear seat passenger compartment.

In addition to its rearward position, the door measures 3.5" wider than comparable coupe doors on previous full size car models. This provides a 37" wide access to the passenger compartment when fully opened.

These features, in combination with the smooth low floor, make the rear seat as accessible as on many four door models.

Headlamps are concealed in the front sheet metal and are raised and lowered automatically when the lamps are turned on. Actuator units are vacuum power cylinders similar to power brake cylinders, with the pistons connected to the headlamp assemblies through linkage. These cylinders produce excellent opening and closing forces with good reliability.

The car uses a new quiet, draft-free ventilation system which eliminates vent windows and exhausts body air beneath the backlight. Air enters the car through inlets in the cowl sides and instrument panel. It then flows over the rear seat passengers and out beneath the rear seat. From there it passes into a plenum, through the one way valve and out through the grille beneath the backlight.

Ventilation air inlets in the cowl side and instrument panel provide ideal distribution of air. The upper inlets in the instrument panel are fed by a blower which gives ventilation at low car speeds when ram air is not effective.

A great deal of careful work was done to develop a superior air exhaust system which provides automatic operation with ideal flow over rear seat passengers and with no noise feedback or possibility of air backflow from the rear outlet. The secret to this accomplishment is the pressure operated one-way valve ahead of the outlet grille.

Elimination of the vent windows provides a quieter car for wind noise with windows either open or closed. Further, when the front window is partially opened, air is drawn from the car the same as with a vent window but with less noise.

A lot of thought and effort went into the design of the car's interior. The car was asking for an approach to interior design that got away from the typical floor console over the tunnel and made maximum use of the better seating and flat floor.

The result is the new Strato-Bench Seat which combines the sportiness and support of a bucket seat with three passenger utility.

The instrument panel was designed with controls concentrated in front of the driver, and the panel curves away from front seat passengers to increase front compartment room. In effect, the traditional floor console has been placed in front of the driver where he can use it.

That covers the arrangement of car components. The car's exterior appearance is designed to reflect honestly this mechanical arrangement. The long, powerful hood, low body silhouette, and sleek, tapering rear end reflect the arrangement of car components. The aim of this functional design approach is to give a feeling of rightness to the car's appearance. The result is styling that makes the vehicle's design aspects strongly apparent.

We made one very important decision at the beginning of the development program on the car, and that was that no compromise would be made on car handling. Initial experience indicated that the design had inherently good handling, and we decided to concentrate to our utmost on enhancing this quality.

The result is a car with good directional control characteristics at all speeds and almost complete freedom from wander during highway driving in gusty crosswinds. We think it represents a real step forward in the area of car behavior.

An important factor in the car's handling and lack of roll are the high spring rates used. Front wheel rates are 89% higher and rear rates are 57% higher than a comparable weight rear drive Oldsmobile. It is interesting how well this car accepts these spring rates. The ride is surprisingly good for the rates used.

In addition to the beneficial effects on handling and roll control, another advantage in the high spring rate combination is the relatively small change in car height from unladen to fully laden. This gives good appearance at all passenger loadings and reduces the possibility of the suspension bottoming out.

The Toronado represented a challenging design and development problem. It presented equally challenging problems in Engineering Organization and Management.

As is the case with most GM Car Divisions; the Engineers at Oldsmobile work on all of our cars. That is, the department is not divided into groups who work on certain cars. Design and development is organized according to the major car components; body, chassis, engine, and transmission. These design groups specialize and handle the development of their components for every car in our line. In addition, we have an Advance Design Group working on new design projects of an exploratory nature, and an Experimental Group who handle all testing.

The Toronado began as a project in the Advance Design Group. Their original assignment was to explore various car configurations with the objective of improving vehicle utility by basic changes in the traditional arrangement of components.

This work progressed well through evaluations of a number of proposals, eventual concentration on front wheel drive and development of the optimum front drive arrangement.

Later, when the decision was made to proceed with the car for 1966 introduction, the Toronado was turned over to the regular design and development groups to handle the intensive work required to ready the car for production.

At that time we considered the possibility of maintaining the Toronado as a separate project and pulling specialists from the various groups to work on the car, the thought being that progress could be better followed with the work under the control of one relatively small group of people. This approach was eventually rejected because it was too limiting. We needed to tap the creative abilities of the entire organization.

We reaped another advantage from this approach to the work. Our Engineering organization and the individuals within it are stronger for having participated in the work from the point of view of pride of accomplishment, as well as the development of their capabilities as engineers.

The method of organizing the work and the efforts and talents of our people were the major factors in the successful development of the car. However, extensive use was made of two essential elements, without which the job could not have been accomplished. These are the computer and accelerated laboratory testing.

These tools are becoming increasingly important to the operation of any modern large scale engineering activity because they enable rescaling time. Both permit accomplishing in hours what used to take weeks.

We use the computer for many tasks from calculating drive train stress to maintaining our critical path follow-up system and issuing parts lists.

Every piece and component of this completely new car had to be visualized, designed and tested in a relatively short period of time. This made us place more emphasis than ever before on accelerated testing in the laboratory.

Laboratory tests were set up in advance for almost every new part in the car during the initial design stage, allowing us to build our test cars with parts almost certain to pass the more time consuming Proving Grounds durability tests. Further, they permitted quick solution to problems that would have taken months if worked out by a series of runs on test cars.

In addition, the greater number of test samples we were able to run increased our understanding of such things as suspension and drive system durability, this covering adequately the full spread of fatigue life from maximum to minimum.

As an example, perhaps the most important phase of the laboratory test program centered around the front wheel drive components themselves, engine, transmission, differential, and drive shaft. To test this complete assembly, a method was developed to evaluate quickly the durability of the design in advance of building test cars.

Testing all drive train components simultaneously was an ambitious project in laboratory evaluation, but the compactness of the front wheel drive system lent itself ideally to this approach. A complete assembly was connected through the front drive shafts to two absorbing dynamometers. An engine provided the driving power. The key to this test was our extensive experience in magnetic tape programming, which has proved itself a favorable tool in engine and transmission development.

To simulate actual driving durability schedules, magnetic tape on a multiple channel programmer controls engine speed, throttle opening, transmission driving range, and individual front wheel speeds. The control tape was made on a vehicle driven over the durability course at the Proving Grounds. When installed on the programmer in the laboratory it repeated the schedule exactly, for over 900 miles a day. This had the advantage of rapid evaluation because it eliminated maintenance down-time, permitted exact reproduction without weather and human variable, and allowed visual evaluation of the components while the test was in progress.

The final phase of the Toronado program involved the building of thirty-seven pilot production cars. These cars were completed in May, 1965, giving our manufacturing people advance experience in assembling the car and giving us three months of final test experience on production cars. The result has been a well organized production start-up.

The reward of all this effort is a really distinctive car with superior mechanical features and behavior that has managed to surpass our greatest expectations. What started out as an interesting engineering project has become reality, a new breed of automobile.

We believe the Toronado represents Engineering achievement, but much more important, it has that magic. It's a great car.

TABLE 1
TABULATION — MAJOR SPECIFICATIONS TORONADO VS. STARFIRE

CAR SPECIFICATIONS

	Toronado	Starfire Coupe
Overall Length	211"	217"
Overall Width	78.5"	80"
Overall Height	52.8"	54.1"
Wheelbase	119"	123"
Tread - Front	63.5"	62.5"
Tread - Rear	63"	63"
Curb Weight	4496 lb.	4346 lb.
Weight-Front	60%	53%
Weight-Rear	40%	47%
Engine Displacement	425 cu. in.	425 cu. in.

Final Form of the 1966 Toronado, as translated into steel, was this six-passenger hardtop coupe, with a modified, 425-cubic inch Toronado V8 with Turbo Hydra-Matic transmission.

The Toronado Takes Shape

From preliminary sketches to final design, the styling evolution of Oldsmobile's front wheel drive Toronado was an extraordinary adventure in automotive architecture.

The stylists who designed the outward appearances and interior comforts of the Toronado tackled a challenge as arduous and exciting as the one Oldsmobile engineers faced in developing the concept of front wheel drive.

No longer do stylists merely "decorate" the end result of hours of automotive engineering development. Nowadays, designing an automobile – especially one as new as the Toronado – is as technical as it is artistic.

The transformation of the Toronado from sketches to sheet metal began early in 1962 at the Oldsmobile studio at GM's Styling Center in Warren, Michigan. An Oldsmobile design team, having all but wrapped up the 1964 Oldsmobiles, was turned loose on a fancy-free assignment to design a dream car, a "car of the future."

In the beginning, it was an exercise in design experimentation, an approach which often leads to the spawning of new automotive styling.

A life-sized, flame-red airbrush illustration of a low four-passenger car was the result. The car constituted something entirely new for the Oldsmobile character and image.

That same summer, Oldsmobile coincidentally sent engineering layouts of its front wheel drive prototype to GM Styling, and the flame-red illustration took on added significance. The design seemed ideally suited to the concepts of the new car.

"The illustration was of a car whose driver would look at home in a Homburg or helmet," recalls one stylist, "and that seemed to fit the basic idea of the new front drive concept perfectly."

Moving swiftly, GM Styling prepared full-sized drawings to study mechanical requirements of the car. Seating mock-ups were built to check the car's various dimensions. The project was an exhilarating one for engineers and designers alike, because front wheel drive offers design proportions and styling freedoms rarely experienced in conventional cars.

While advanced ideas appear sound on paper, it remains for a three-dimensional examination to determine their final merit. Skilled sculptors set to work building a clay model of the unborn car. Working with micrometer precision, the sculptors prepared a clay model Toronado complete with actual wheels and tires and detail hardware, such as door handles and headlights. Plastic sheets simulated glass areas, and metal foil was used to create the chrome and bright work on the car's body.

On a cold day in February, 1963, the clay model was moved outdoors at the vast GM Technical Center in suburban Warren for evaluation.

Management's reaction was one of enthusiasm. A further "go-ahead" was issued.

Wind tunnel tests to check aerodynamics were the next step, and a

Styling birth of Oldsmobile's Toronado began with artist's preliminary sketches.

one-quarter scale model was built for that purpose. The tests prompted the decision, among others, to eliminate the Toronado's corner vent windows.

Interior styling of the car began that spring. A "driver podium" instrument panel was conceived with a vertically revolving speedometer. Arm reach convenience was measured and checked. By January, 1964, the interior was modeled in clay, shown to management, and given tentative approval.

Meantime, the full-sized clay model was continuously being refined and altered, both to meet production requirements and to reflect subsequent engineering modifications as well as styling improvements.

At last, when management issued a full approval of the clay model Toronado, casting in plaster was ordered.

From these molds, GM Styling's fabrication shop made fiberglass panels. Mounted on a dummy chassis with special frames, these rigid panels were painted, chromed, and made to appear real in every detail.

The first fiberglass model was reviewed by Oldsmobile and corporate management in mid-1963. Various changes and revisions were ordered. Finally, 8 months later, GM's board of directors viewed a hand-built version of the car and gave it their official sanction, ordering it into production with the start of the 1966 model year.

Apart from its many new features and its outstanding road performance, Oldsmobile's Toronado is significant as an example of the infinite step-by-step procedure required to design and develop a new automobile.

It also illustrates how deeply rooted tomorrow's automotive styling is in today's technological advancement.

Seven years in development (nearly three devoted to styling), and the Oldsmobile Toronado is an extraordinary car, an adventure from drawing board to driveway.

o

Numerous "dream" car drawings gave inspiration to Toronado's final design.

Painstaking attention to detail results in micrometer precision in the clay model.

Above: Life-sized, airbrush illustration of unborn Toronado gave stylists a pattern.

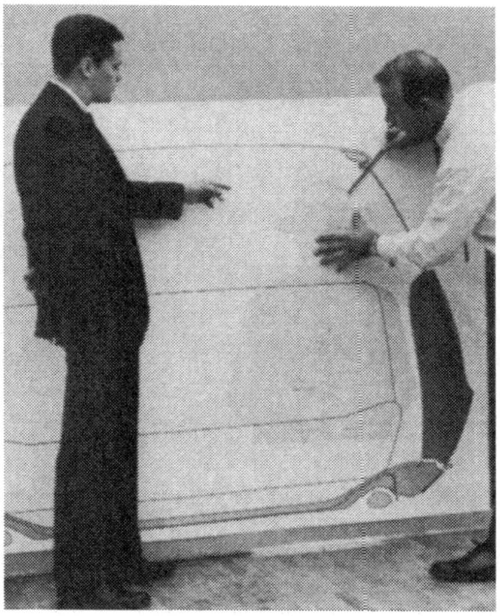

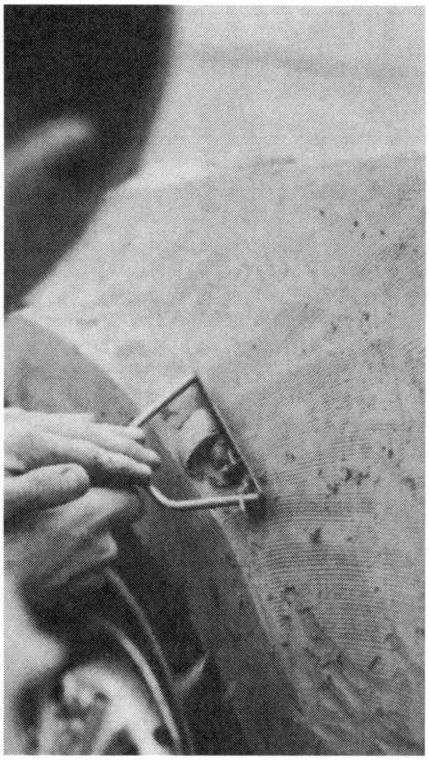

Below: Using airbrush illustration as a guide, skilled model makers build a clay Toronado.

Above: Interior measurements are checked and double-checked for driver and passenger comfort.

Preliminary clay model won enthusiastic approval, and ultimately led to 1966 production order for Toronado.

Technical Report on Olds Toronado

Lansing — Oldsmobile engineers, scoring a breakthrough in front wheel drive design, have developed the Toronado, the most unique American automobile in many years.

To be produced as a six-passenger hardtop coupe, the Toronado features unusual interior space for passengers and luggage, outstanding driving traction and handling characteristics, and styling like no other car on the road. It is 211 inches long, only 52.8 inches high, and is built on a 119-inch wheelbase.

Toronado design objectives have been achieved through the elimination of the transmission hump and driveline tunnel, the development of an advanced torsion bar suspension system, and the location of all power train components under the hood, well forward of the passenger compartment.

The engine is situated as usual fore and aft under the hood, but is offset slightly to the right of center. The rest of the power train design, through which torque is transferred to the front wheels, is ingenious.

The transmission's torque converter is attached conventionally to the rear of the engine. The remaining seciton of the Turbo Hydra-Matic transmission is turned around and mounted along the left side of the engine, facing forward. A link chain assembly transfers power from the converter across the through the transmission gear sets to a differential unit bolted to the front of the transmission, which in turn splits the torque between two front drive axles.

Its power plant, called the Toronado V8, is a modified, more powerful version of Oldsmobile's 425-cubic inch engine, and is rated at 385 horsepower. Performance increases result from improved engine breathing and completely new, more efficient carburetion.

A new four-barrel carburetor, the Quadrajet, developed for Oldsmobile by Rochester Products division, increases air induction. Primary stages are 22% smaller, using a triple venturi for finer, more stable mixture control providing improved idling and increased economy. Secondaries are 44% larger, also utilizing more precise fuel control called "air valve metering," which insures correct air/fuel ratios during acceleration and high-performance operation. The new, more simplified design uses a single fuel reservoir with just one inlet needle and one float to assure a constant fuel supply in any driving situation.

Overall engine operation is improved through a new choke which is mounted in the intake manifold, where it can better sense engine temperatures.

A large, low profile, dual snorkel air cleaner with a resin treated filtering element supplies air to the Quadrajet. In addition to furnishing air, the tapered snorkels also contribute to the suppression of noise, while their efficient, thin design allows for a lower hood.

Larger intake valves, an increased throat dimension in the head and larger, less restrictive intake manifold branch areas, provide improved engine breathing for more performance. Toronado engine output is also increased by a high-lift profile camshaft and larger valve lifters.

Teamed with the Toronado V8 engine as standard equipment is the specially adapted version of Oldsmobile's Turbo Hydra-Matic transmission. In function and controls, it is basically the same as the standard three-speed Turbo Hydra-Matic, but in construction and power transfer, it is completely new.

Transmission power flow has to be reversed so that driving torque can be applied to the front wheels. This is accomplished by dividing the standard Turbo Hydra-Matic into two units.

The variable stator torque converter is attached conventionally to the rear of the engine. However, the gear sets and controlling elements are rotated 180 degrees, are enclosed alongside the converter, and extend forward along the left side of the engine. A 2-inch wide, multiple link chain rotates counterclockwise, transferring torque from a drive sprocket at the rear of the converter to a driven sprocket at the rear of the gear sets.

Instead of a long drive shaft running back the entire length of the car underbody, a simple spline shaft connects transmission gear output with the differential. Eight bolts fasten the differential directly to the front of the transmission.

The Toronado's differential has been specifically designed for this car. It uses a planetary gear set rather than the standard ring gear-pinion, allowing for its slender, more compact design. Differential torque is split between two front drive axles; one is connected to the planetary gears, and the other to an internal sun gear.

While the car is moving straight ahead, gears are fixed and rotate at the same speed, but when turning, the planetary gears revolve with differential action, allowing the drive axles to rotate at different speeds.

Torque is fed from the differential directly to the left drive shaft, and through a cross shaft under the engine oil pan over to the right drive shaft. Each axle drive shaft has inboard and outboard constant velocity universal joints so the shafts have upward, downward, forward, or rearward freedom required for independent front suspension.

Front suspension for the Toronado is of torsion bar design. Two specially hardened steel torsion rods work from each lower control arm to special anchors mounted in a cross support near the middle of the car. The cross support, mounted with rubber cushions to the frame, contains single bolt anchor adjustments whereby the carrying height can be adjusted by increasing or decreasing torque on the torsion rod.

Both upper and lower control arms are heavy stampings and specially calibrated shock absorbers for fast reacting stability are mounted at an angle. Wheel caster and camber can be easily adjusted with an eccentric arm assembly in the upper control arm, rather than by the conventional shim method.

A rugged stabilizer bar across the fron between the lower control arms, keeps Toronado's front end flat and level when cornering.

Integrally designed steering knuckles and plane arms are simplified one-piece forgings.

Standard power steering for the Toronado is responsive and precise with an overall ratio of 17.8 to 1. Lock-to-lock is about 3½ turns, compared to nearly 5 turns for conventional steering. Proper steering column angle is achieved with a flexible, constant speed universal joint that also isolates vibrations.

A special shock absorber mounted from the frame cross member at the front to the steering linkage intermediate rod, adds to the car's outstanding steering control characteristics.

Supporting front suspension and power train components is a perimeter "stub" frame terminating with a rear cross bar that supports forward eyes for the rear springs. Box section construction is used for better beaming and torsional rigidity.

Rear suspension consists of a stamped U-channel axle with "dead" spindles bolted on for the rear wheels. The axle is cushioned by two single leaf springs and four shock absorbers. The leaf springs are anchored with rubber bushings and are fastened with shackles to an integral underbody frame at the rear.

Two horizontal and two vertical shock absorbers are used to dampen rear spring windup, for maximum controlled braking, for the elimination of wheel hop, and generally, for a much smoother, quieter ride. Oldsmobile's Toronado is the only automobile to use four rear shock absorbers.

A new brake drum has been designed for the Toronado with cast fins providing quick heat dissipation for improved brake life and reduced fade characteristics. Enlarged brake linings (11 x 2¾-inch front and 11 x 2-inch rear) are used with self-adjusting primary and secondary shoes.

Impressive looking 15-inch wheel spiders have functional openings offset into the air stream, so that the air flows through the openings and over the cast brake drum fins. Tires are 8.85 x 15.

An interesting mechanical innovation for the Toronado is a new self-equalizing accelerator cable that operates the carburetor. The simple throttle device includes a spring-loaded downshift detent to retain the "kick-in" feel when passing.

Another new feature in the Toronado is a quiet, draft-free ventilation system that eliminates corner vent windows and exhausts air under the rear seat and up through louvered outlets below the back window. This assures a quieter passenger compartment and reduces wind noise. When the side window is partially opened, air is drawn from the car in the same manner as with a corner vent, with less noise.

Air inlets in the Toronado's cowl sides and instrument panel provide more uniform air distribution. The upper inlets in the instrument panel are blower fed, enabling comfortable ventilation at low car speeds when ram air is less effective.

Still another Toronado feature is its retractable head lamp system which is completely automatic, operating entirely from engine vacuum. When the head lamp switch is pulled on, the lamps automatically raise to the open position in seconds. Pushing the switch to the park or off position automatically retracts the lamps. A large vacuum reserve tank (600 cubic inches) insures three full cycles with the engine off. Parking lamps operate independently.

o

1966 OLDSMOBILE SPECIFICATIONS – STARFIRE, 98, TORONADO

	Starfire	98 (1)	Toronado
Overall Length	217.0	222.9	211.0
Overall Width	80.0	80.0	78.5
Overall Height	54.1	55.8	52.8
Wheelbase	123.0	126.0	119.0
Tread			
Front	62.5	62.5	63.5
Rear	63.0	63.0	63.0
Tires			
Standard	8.25 x 14	8.55 x 14	8.85 x 15
Optional	8.55 x 14	8.85 x 14	9.15 x 15
Brake Dia. & Lining Width			
Front	11 x 2.75	11 x 2.75	11 x 2.75
Rear	11 x 2.00	11 x 2.50	11 x 2.00
Total Brake Lining Area	208.6	229.2	208.6
Weight			
Shipping	4013	4177	4311
Curb	4204	4368	4496
Engine	V8	V8	V8
Maximum Brake Horsepower	375	365	385
Maximum Torque	470	470	475
Displacement	425	425	425
Bore & Stroke	4.125 x 3.975	4.125 x 3.975	4.125 x 3.975
Compression Ratio	10.5	10.25	10.5
Carburetion	4 bbl.	4 bbl.	4 bbl.
Crankcase Capacity	4 qts.	4 qts.	5 qts.
with Filter	5 qts.	5 qts.	6 qts.
Electrical System	12 volts	12 volts	12 volts
Cooling System Capacity	16.5 qts.	16.5 qts.	16.5 qts.
with Heater	17.5 qts.	17.5 qts.	17.5 qts.
Fuel Tank Capacity	25 gals.	25 gals.	24 gals.
Transmission			
3-speed Fully Synchronized	Std.	N.A.	N.A.
4-speed Fully Synchronized	Opt.	N.A.	N.A.
Turbo Hydra-Matic	Opt.	Std.	Std.
Rear Axle Gear Ratios Std. (2)			
w/3-speed	3.23	N.A.	N.A.
w/4-speed	3.23	N.A.	N.A.
w/Automatic	3.08	3.08	3.21
Turning Diameter	43.6	44.5	43.0
Trunk Capacity	19.4 cu. ft.	23.4 cu. ft.	14.5 cu. ft.

(1) The Starfire engine (375 hp) is available at extra cost on all 98 models.
(2) Other rear axle ratios available for special requirements.

Unusual grille design invariably drew comment about equally divided between favorable and unfavorable. Headlights are fixed and the cornering lights are set low in the fenders.

Pull vs. Push

Some owners forget and put chains on the rear wheels of their front-drive Toronado!

Introduced in 1966 with a shape vaguely reminiscent of a purportedly female Soviet discus hurler, the Olds Toronado never made much of a sales mark until its image was changed from muscle to svelte in 1971. The '72 model tested here is essentially the same except for a switch from cross-hatch to vertical grille bars.

The muscle remains but it's hidden under rather attractive sheet metal with some unusual styling features. Chief among them is an extra set of tail and brake lights mounted just under the rear window. There's no mistaking a Toronado, particularly at night in heavy traffic. Least among them is a sort of gimmicky outside air temperature indicator that may be optionally mounted adjacent to the driver's side-view mirror.

One of the big changes that came with the new styling was a full-length frame with coil spring rear suspension. Previous versions had an abbreviated frame stretching to a point just short of the rear axle and leaf springs that were first single and then multiple semi-elliptics. It's been a while since we've driven one of the older cars but our impression is that they had a better, more stable ride.

Torsion bars are retained in front as is the unique chain-belt drive to the Turbo Hydra-Matic mounted to the left of the engine. Only the planetary gearset is offset, the torque converter remaining in its usual position at the rear of the engine. Another unusual feature is the longer right-hand half shaft passing through a cleft in the oil pan. This drivetrain design is now in its sixth year of production at Oldsmobile, has been adopted almost intact in the Cadillac Eldorado, and to our knowledge, has caused no chronic problems.

It probably is true that some Toronado owners don't even realize that their car drives through the front wheels. A friend who operates a service station in the Lake Tahoe area confirms this. He has had Toronado customers who insist on having chains and/or snow tires installed on the rear wheels!

Certainly, the average driver couldn't tell from the way the car handles. No special techniques are required for cornering under normal conditions though

Toronado avoids the controversial "boattail" look of Buick's Riviera with which it also shares body shells. Second set of taillights are recessed under the back window.

Overall length is 220 inches, three shorter than the Eldorado, on a wheelbase of 122 inches. Rear quarter windows, optionally electric, slide into quarter panels.

Instrument panel has a rather barren appearance for a luxury car. Thick wheel spokes can block view of speedometer in certain positions with optional tilt and telescope column.

like the Eldorado, you can get around fastest using a combination of brake and throttle. The Toronado is more stiffly sprung than its more expensive counterpart which gives it an edge in agility, mainly an ability to stay within its own lane on a curve, but the springs and shock absorbers have a problem cooperating. You get annoying rebound when crossing railroad tracks and the like. If the bump or dip happens to be on a curve, the temporary loss of traction on the steering wheels causes the whole front end to move over a foot or two. Your natural reaction is to apply more steer and that only aggravates the situation.

These "Supershocks," as Oldsmobile calls them, have Teflon coated pistons and valves which are impervious to moisture. A plastic envelope filled with Freon is immersed in the hydraulic fluid and it's supposed to de-aerate the mechanism. Whether it does or doesn't, those Supershocks are slow on the draw. As with the Eldorado (RT, April '72), we'd switch to Konis or some other brand of adjustable shock if we owned the car.

Another less than desirable feature of the ride is sensitivity to the symmetric undulations characteristic of heavily traveled urban freeways. Our test Toronado reacted to these as though all four tires had suddenly developed flat spots. The Eldorado driven at the same speed over the same route was not troubled at all.

The instrument panel is one of GM's newer designs which dismantles easily for servicing. Switches and the clock snap in place and the lower valance panels can be removed to permit relatively easy access to warning light bulbs and the speedometer. Esthetically, however, the layout is not too inspiring as the only gauges provided are the square speedometer and rectangular gas supply indicator. An optional, overly bright gas warning light glares at you continuously whenever the fuel supply dips below the quarter mark which seemed to be most of the time with this thirsty car.

The standard 455 cu in. V8 (the only engine offered) lost another 25 horsepower this year, dropping to 250 SAE net at 4,000 rpm due to additional emission control equipment and settings. It's a smooth engine well endowed with low-end torque which ac-

With a 63.5-inch track and fairly stiff springing, the Toronado corners well at speed. Ride, though, is troubled by secondary jounce from shocks.

counts for the rather excellent acceleration times we recorded. The standing quarter in 17.2 seconds at 84 mph is quite respectable for a 4,660 lb car with a 3.08 differential. Guzzling gas in traffic at the rate of 7.2 mpg, on the other hand, effectively eliminates any desire to make use of the performance potential.

An exclusive and desirable feature of Oldsmobile engines is positive rotation for both the intake and exhaust valves. Rotation was originally developed by the Ethyl Corporation to alleviate valve burning caused by unequal mixture distribution in long straight-eight engine blocks. In the more compact V8 shape, distribution is not much of a problem but Olds uses rotation to prolong valve life and thus new car standards of performance and emission control. The wiping action of the valve face against the seat maintains a clean, precise contact.

A problem with '71 Toronados overheating has been cured by mounting the fan shroud on the engine so that a minimal 0.2 in. clearance may be maintained with no risk of the fan scraping the shroud as it would if the latter did not move with the engine. The closer the shroud is to the blade tips, the more effective is the cooling action except at high car speeds. For that reason, a portion of the radiator surface is left unshrouded.

All Toronados have disc front brakes with power assist as standard equipment but new this year is a brake pad wear sensor. It consists of a small protruding metal finger mounted on the caliper that scrapes noisily on the disc when pad wear goes beyond safe limits. The scraper does not mark the disc in the area contacted by the pad and it effectively reminds you to replace the pads before they do enough damage to require an expensive disc regrind.

Our test car was not equipped with the optional "True-Track" anti-skid system and brake performance was superior to that of last month's test Eldorado which was so equipped. The difference was not so much in distance from 60 mph but in freedom from fade. Six repeated maximum stops did not produce any increase in pedal pressure although the rear drums did tend to lock up toward the conclusion of each stop.

Considering the $106 difference in initial cost, we're not too sure that the more expensive automatic "comfortron" air conditioning system is a wise investment. The manually controlled "Four-Season" unit (at $416.80) was easy to fine tune for conditions we encountered although admittedly, they were not extreme. The ill-conceived vents in the trunk lid have been abandoned this year in favor of outlets in the door lock pillars to preclude any possibility of a negative pressure condition that sucked exhaust gases into the car. By mid-1971, all GM senior cars were emerging from the factory with the old style vents half blocked which reduced efficiency, and earlier production cars were retrofitted by dealers.

If you favor fresh air, power windows are a worthwhile option at $129. The rear quarter windows slide neatly into the landau area of the roof while front windows may be opened partially or fully without too much draft entering the car. Both the window controls and those for the optional six-way seat are located in the forward part of the armrest along with the lever for the remote mirror. We enjoyed not having to fumble under the seat to change its position but we also wondered why they couldn't have put the power door lock switch in the same grouping. It has its separate position down alongside the door handle and thus, we seldom thought to use it.

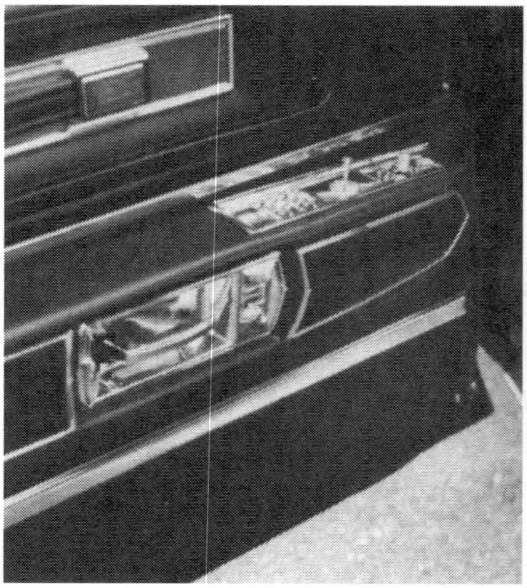

Placing both seat and window controls on the armrest is a real convenience but the door lock control to the right of the handle could be better placed.

A Toronado owner would be well advised to keep an extra set of keys in his pocket at all times because once the electric lock is set, you just slam the doors. This happened to us in a service station when we left the car to wash up with the keys still in the ignition. While

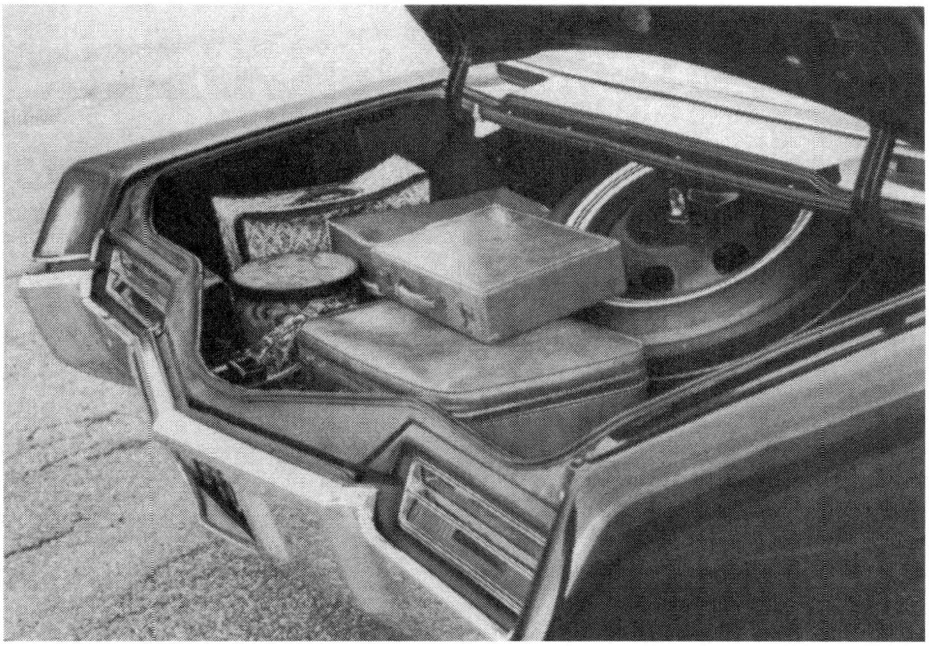

Braking is even and fade-free with the rear wheels tending to lock up toward the end. On test car, transmission popped into neutral during each stop.

Spare occupies a lot of valuable space in the smallish, 13.5 cu ft trunk. An inexpensive electric lock operated from the glove compartment is optional.

reaching for the inside hood release, the attendant inadvertently contacted the lock switch and then closed the door. A considerable delay ensued before a locksmith could be contacted to get the car open again.

Rather than being crowded in among other switches, the windshield wiper and washer controls are at the far left of the panel where they may be reached without having to look for them. In addition to the usual two speeds, the wipers had a "mist" position which when flicked, would cause them to operate for a single cycle. Or, you could hold the switch for more sweeps. This proved quite useful during the foggy conditions that plagued most of our week with the Toronado.

The tilt and telescope feature of the steering column was all but negated by the fat spokes of the wheel itself. With the wheel pulled out or tilted up, you can't read the speedometer or gas gauge, much less the position of the shift lever. You also can't use those spokes for a hand rest as very little pressure will actuate the horn.

Mention of the shift indicator reminds us that the transmission invariably popped into neutral after each of our maximum brake stops. Just at what point this happened, we don't know as there was no accompanying noise but obviously, the braking effect of the engine was lost. Very probably, this was a fault peculiar to our test car as we haven't encountered it with Toronados we've driven in the past.

Another thing we learned is why you seldom see a black car of any make nowadays. Ours was black inside and out and it was all but impossible to keep clean. Other color cars undoubtedly get just as dirty but you don't notice it. Also sometime, try untangling black seat belts on a black seat at night, or finding a black wallet that slipped out of your pocket.

As the current Toronado body shell is only in its second year of usage, it features most of the latest safety innovations such as inner and outer roof panels and impact guards in the doors and rear quarter panels. It doesn't, though, have the sprung bumpers introduced by Oldsmobile on '72 rear-drive cars. These are not the 5 mph safety bumpers soon to be required by law but they are capable of flexing up to one and one-half inches and thus escape damage from the usual parking lot encounters. A virtue of the carryover Toronado bumper, however, is that it is assembled in six sections that may be individually replaced in the event of damage.

Just as with the Eldorado, relatively rapid front tire wear is a problem with which Toronado owners must live. Rotation every 6,000 miles is economical and recommended, following a pattern of rear to opposite front, front to the same side rear and not using the spare. Inflation pressures as posted on the glove box door should be checked weekly as J78 x 15s of good quality cost at least $50 each. With this and rotation, you can do much to avoid the excessive wear reported by owners who don't maintain their tires. Put another way, Michelin confirmed to us that their 40,000-mile warranty covers the Toronado if the car and the tires have been properly serviced.

If front drive is the primary feature of interest in this size car you have, of course, only the Toronado and Eldorado to choose between unless you are attracted by the even more expensive Citroen SM. While we'll grant that Cadillacs probably receive more care in assembly, the materials used in the two cars are generally of the same quality. In base form, the Toronado costs $5305.60 versus $7383.00 for the Eldorado. In test form, the Toronado listed at $7393.80 versus $10,140.35 for the Eldorado.

Those figures speak for themselves. Due to the interchangeability of the body shells, the two cars have practically identical interior dimensions. Both perform about equally. The Eldorado has a slight edge in luxury but that gets to the point of no return. It boils down to whether you'd rather have an air-conditioned, completely equipped Toronado or a bare bones Eldorado without even a radio for the same dollars. Or, perhaps, the decision is just how much are you willing to pay for that Cadillac name plate? ●

Fan shroud is mounted directly to the engine to allow minimal fan blade clearance and connected to the radiator by a flexible shroud extension.

OLDSMOBILE TORONADO 2-DOOR HARDTOP

SPECIFICATIONS AS TESTED

Engine	455 cu in., OHV-4V V8
Bore & stroke	4.125 x 4.250 ins.
Compression ratio	8.5 to one
Horsepower	250 (SAE net) at 4000 rpm
Torque	375 lbs-ft at 2800 rpm
Transmission	3-speed, Turbo Hydra-Matic
*Steering	3.5 turns, lock to lock 44.9 ft, curb to curb
*Brakes	disc front, drum rear
Suspension	torsion front, coil rear
Tires	J78 x 15, belted bias-ply

Dimensions (ins.):

Wheelbase	122.0	Rear track	63.6
Length	220.6	Ground clearance	5.0
Width	79.8	Height	54.7
Front track	63.5	Weight	4660 lbs

Capacities:

Fuel	24 gals	Oil	6 qts
Coolant	19.5 qts	Trunk	13.5 cu ft

*Power assisted as tested

BASE PRICE OF CAR

(Excludes state and local taxes, license, dealer preparation and domestic transportation): $5306 at Lansing, Mich.

Plus desirable options:

$ 431 Air-conditioning
$ 233 AM/FM stereo radio
$ 135 Vinyl roof
$ 180 6-way seats (both sides)
$ 129 Electric windows
$ 154 Custom trim
$6568 TOTAL

ANTICIPATED DEPRECIATION

(Based on current Kelley Blue Book, previous equivalent model): $728 1st yr. + $1186 2nd yr.

PERFORMANCE AND MAINTENANCE

Acceleration:	Gears:
0-30 mph	3.9 secs., 1st
0-45 mph	6.4 secs, 1st, 2nd
0-60 mph	10.0 secs., 1st, 2nd
0-75 mph	14.9 secs., 1st, 2nd
0-¼ mile	17.2 secs. at 84 mph
Ideal cruise	75 mph
Top speed (est)	115 mph
Stop from 60 mph	162 ft
Average economy (city)	7.2 mpg
Average economy (country)	11.8 mpg
Fuel required	91+ Octane
Oil change (mos./miles)	4/6000
Lubrication (mos/miles)	4/6000
Warranty (mos./miles)	12/12,000
Type tools required	SAE
U.S. dealers	3290 total

RATING	Excellent (91-100)	Good (81-90)	Fair (71-80)	Poor (60-70)
Brakes	91			
Comfort		88		
Cornering		90		
Details		89		
Finish	93			
Instruments			80	
Luggage		82		
Performance	93			
Quietness	91			
Ride		85		
Room		90		
Steering	91			
Visibility		88		
Overall		88		

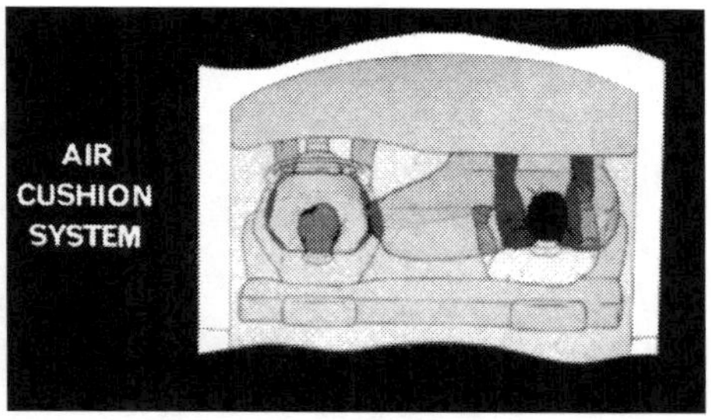

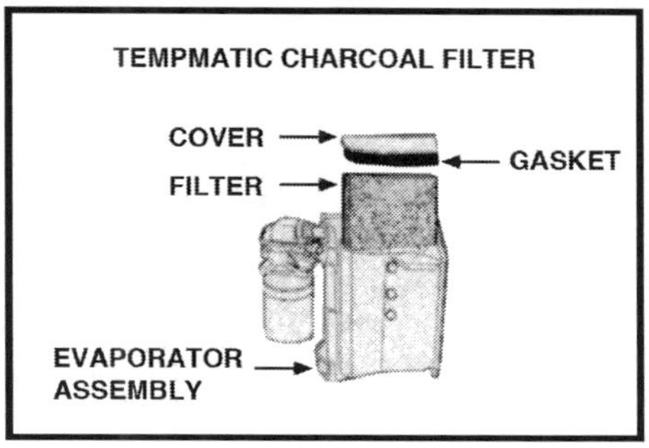

Oldsmobile Toronado

(Custom Coupe, Brougham Coupe)

Still Riding The Crest

One of America's two front-wheel-drive automobiles is the Oldsmobile Tornado. Fitting the personal luxury car slot, the Toronado has a minor front end redesign and a new model, the Brougham Coupe, which is basically a decor option rather than a technical change.

Both the Custom Coupe and Brougham are built on the 122-inch wheelbased chassis shared with Cadillac's front-drive Eldorado. The Toronados use the 455 CID V8 engine and Hydra-matic transmission. The rear axle ratio is 2.73:1 standard and a 3.07:1 optional.

Improved handling can be attained by using the GM Specification radial-ply tires. These new tires are what GM feels is the best possible combination of ride, handling and tread life for its cars. GM set the criteria and the major tire manufacturers conformed.

One feature pioneered on the Toronado has been adopted by other Olds cars: the spring-steel tabs that contact the front discs as the brake pads wear down, thus giving an audible signal when it is time to change pads.

The Toronado engine, like the other Olds 455 powerplants, has the high energy ignition system which eliminates the distributor points and condenser while providing a longer duration and hotter spark for improved starting. The big V8 also has a cold air duct connection to improve hot weather driving by lowering the engine temperature.

Oldsmobile has done little to change the Toronado because of the high customer acceptance of this personal luxury car. No doubt, the same popularity will continue in '74.

GENERAL SPECIFICATIONS

Dimensions, Ins: Length	228.0
Width	79.5
Height	53.3
Wheelbase	122.0
Track, front/rear	63.5/63.6
Luggage capacity	17.0 cu ft
Engine	455 CID V-8 (4-bbl)
Optional engine(s)	None
Horsepower	230 at 3800 rpm
Torque	N/A
Transmission	3-speed automatic
Optional transmission(s)	None
Brakes	Power disc/drum
Fuel capacity	26 gal
Fuel economy	6.8 mpg

Base price (excludes state and local taxes, license, dealer preparation and domestic transportation): $5560

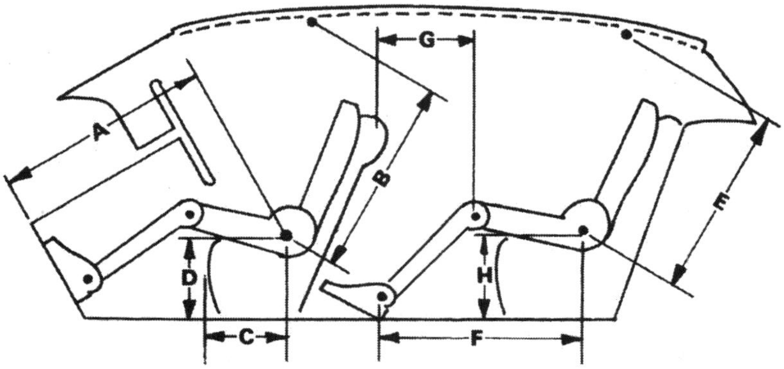

(2-Dr)

A-Front leg room, max, +10"	42.2
B-Front head room, +4"	38.1
C-Front seat travel	N/A
D-Front heel to hip height	N/A
E-Rear head room, +4"	37.1
F-Rear leg room, min, +10"	35.2
G-Rear knee room, min,	N/A
H-Rear heel to hip height	N/A
Front shoulder room	64.3
Rear shoulder room	60.2
Front hip room	62.3
Rear hip room	56.0

MOTOR TREND Road Test

TORONADO, THUNDERBIRD, GRAND PRIX AND RIVIERA

You can get cozy with that "personal luxury car" if you've got $6000 to $8000/By Jim Brokaw

Back in the good old days of mosquitos, Sears' catalogues and frequent flat tires, Ford introduced a magnificent two-seater pseudo-sports car known to all as the T-Bird. In 1958, the Dearborn clay daubers added a rear seat to the 'Bird and it's been growing ever since. Regardless of whether two-and-a-half tons is *over*grown, Thunderbird kicked open a permanent slot in the auto market which tends to draw more competition with the passage of time. The 1973 roster includes Riviera, Toronado and maybe Grand Prix. Also Monte Carlo, Olds Cutlass S, Eldorado and Lincoln Mark IV, if you want to get tacky about it, but we're only testing four of them. The marketing term for this class of conveyance is "personal-luxury." Personal because they used to have distinctive styling, luxury needs no explanation.

Our annual glance at the 'Bird market produced a very pleasant surprise. The intrusion of stability-tuned suspensions and steel-belted radial tires have brought a measure of handling to these symbols of success.

Handling is a relative term not to be confused with handling in the sense of a Capri or a Lotus. With three of the samples lugging around two-and-a-half tons of plastic, rubber and sheet metal, getting around a corner at speed in one piece constitutes handling. All four cars are capable of blitzing moderately exciting turns if done in the correct manner.

To begin with, Thunderbird, mounted on standard steel-belted Michelin radials, heels over moderately on a corner, but once set maintains a steady attitude, returning to level with one solid movement on the straight. Oddly enough there is very little of the plowing normally expected from one of these big boys. Naturally, everyday driving is quite stable. There is one slight clinker though, tire pressure is critical. Our test car didn't have any tire pressure information, but yours will have a booklet in the glove box. We found that 30 psi gives the best handling, but there is penalty of harshness at freeway cruising speeds. The best compromise seems to be 28 psi. When, and if, you purchase any heavy vehicle with steel-belted radials, be sure you get tire pressure information from the dealer before you drive off the lot. It do make a difference.

Toronado is another welcome surprise. The suspension engineers at Oldsmobile have certainly done their homework. For those of you who didn't know, Toronado is a front-wheel-drive machine. With the size and bulk of domestic drive-trains, we naturally expected a heavy weight bias on the front suspension. Not so. Toro's 5110 pounds are split 60% up front and 40% in the rear. Thunderbird divides at 59% in front and 41% at the rear. Not that much difference. Part of the trick is the Toro's powerplant, which sits much further back in the engine compartment than any of its brethren.

There is, of course, a measure of technique required to get the Toro to behave. The time-honored method of backing off the throttle approaching a corner, then applying increasing power throughout the turn is very essential. If you should be more brave than smart and wind up trying to smooth out the wrinkles on the brake pedal going into a turn, the old girl will plow like a draft horse.

Riviera sets a different tone than Thunderbird and Toronado. Suspension is stiffer, giving it more stability. The undercarriage geometry enables it to ignore large bumps at speed, but shows some sensitivity to the short, sharp lumps. By tucking the nose to the inside of a corner, Riviera keeps its head all the way through.

Grand Prix is a bit of an enigma because we're not really sure it belongs in the upper stratosphere of the personal-luxury class. It is 500 pounds lighter than the others, with only a 116-inch wheelbase, compared to 120-plus for the competition. As you may expect, with the short wheelbase and traditional Pontiac attention to springs and things, the Grand Prix handles like the leading lady in a skin flick. All you have to do is point it and punch the throttle. Nimble and quick, the GP must pay the penalty of ride harsh-

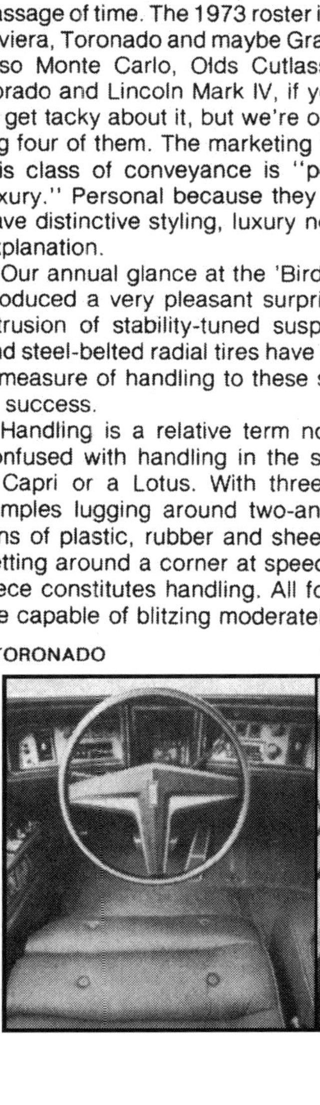

TORONADO

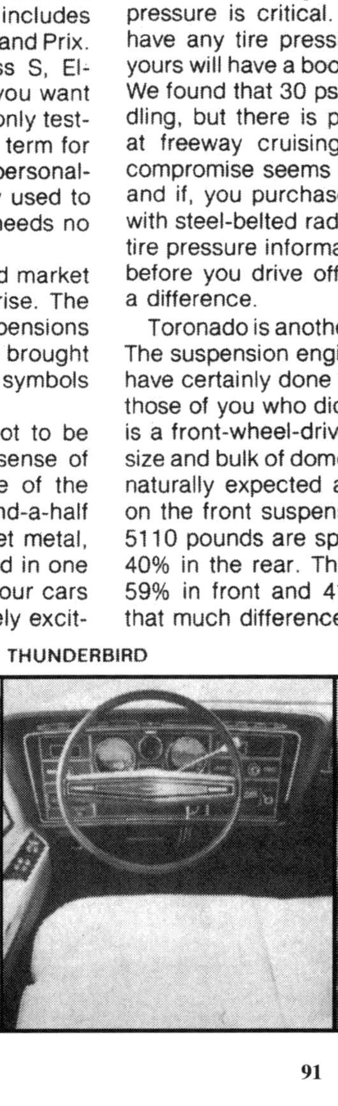

THUNDERBIRD

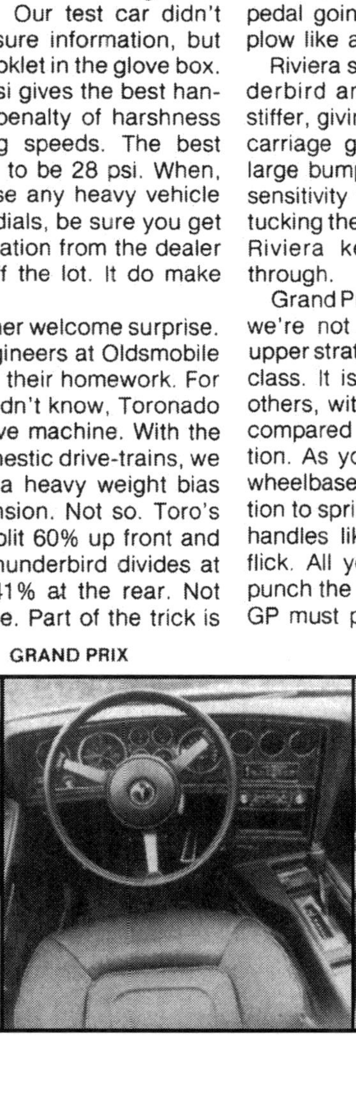

GRAND PRIX

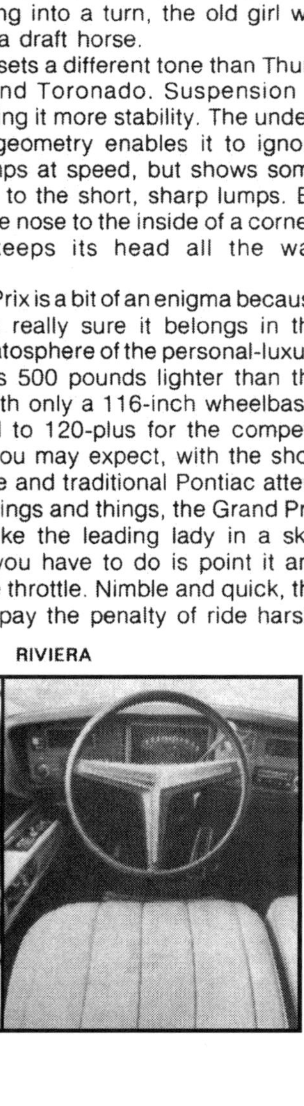

RIVIERA

ness. Relative to the other three of course. Stiffness is not of the kidney destruct variety, but it is a constant companion.

Interior comfort and decor is where it has to happen for these Wall Street sports cars. Seating arrangements are intended to accommodate two forward, however. All save the Grand Prix will permit a third person in front. Thunderbird and Riviera use the split-back bench with individual seat controls. Power controls for the right side are optional extras. Toronado employs a two-one split up front with a small individual seat for the driver and a wide section for passengers. Grand Prix goes the true bucket route with center console. All are six-way adjustable. Three passengers fit easily into all four rear seats.

Both Thunderbird and Riviera omit a specific lumbar support feature, which you may not notice, but I did. Toronado has the most firm seating with the best support. Grand Prix has the least comfortable. GP makes a gallant attempt, but overdoes it with padding under the thighs and too much convex curvature of the seat back center panel. Comfort is very much an individual thing, what pleases me may not please you and vice versa. However, I routinely spend two and a half to three hours a day on the road. Firmness and support count.

All three have the instrument panel clustered in front of the driver, with most of the gauges quite visible. Grand Prix has the best of all with oil pressure, water temperature, fuel and ammeter dials directly in front of the driver.

Riviera has a slight problem in the placement of the transmission selector lever. The right hand strikes the shift lever when reaching for the radio volume control knob. A change of angle on the shifter would help.

Toronado sports a cross-brushed gold dash panel that really gets the job done. Definitely uptown, unless you can't afford a Toronado, then it's crassly ostentatious. Another pleasing feature is the little row of toggle-switch accessory controls mounted along the upper edge of the dash.

Thunderbird's contribution to if-you-got-it-flaunt-it, is their cigarette lighter. It is four inches long with more stages to it than a Saturn V rocket and is more of

MOTOR TREND Test Data

SPECIFICATIONS	GRAND PRIX	THUNDERBIRD	RIVIERA	TORONADO
Engine:	OHV V-8	OHV V-8	OHV V-8	OHV V-8
Bore & Stroke—ins.	4.152 x 4.210	4.362 x 3.850	4.312 x 3.900	4.126 x 4.250
Displacement—cu. in.	455	460	455	455
HP @ RPM	250 @ 4000	208 @ 4400	250 @ 4000	250 @ 4000
Torque: lbs.-ft @ rpm	370 @2800	338 @ 2800	375 @ 2800	375 @ 2800
Compression Ratio	8.0:1	8.0:1	8.5:1	8.5:1
Carburetion	4-V	4-V	4-V	4-V
Transmission	3 spd. auto Torque converter	3 spd auto Torque converter	3 spd auto Torque converter	3 spd auto* Torque converter
Final Drive Ratio	3.08	2.75	2.93	2.73
Steering Type	Variable power	Power	Variable power	Power
Steering Ratio	18.3-14.8:1	21.73:1	16.6-14.56:1	17.9:1
Turning Diameter (curb-to-curb-ft.)	39.9	43.0	41.7	45.65
Wheel Turns (lock-to-lock)	3.5	3.99	2.94	3.24
Tire Size	GR70-15	230-15X	HR70-15	J78-15
Brakes	Power disc/drum	Power disc/drum	Power disc/drum	Power disc/drum
Front Suspension	Coil springs stabilizer shocks	Coil springs stabilizer shocks axial strut	Coil springs stabilizer shocks	Torsion bar/ stabilizer shocks
Rear Suspension	Pivoted 4-link coil/shocks stabilizer	Coil/shocks stabilizer	4-link coil shocks	4-link coil/ shocks
Body/Frame Construction	Separate frame	Separate frame	Separate frame	Separate frame
Wheelbase—ins.	116.0	120.4	122.0	122.0
Overall Length—ins.	216.6	218.90	223.4	226.8
Width—ins.	78.7	79.7	79.9	79.8
Height—ins.	52.9	53.07	54.0	53.2
Front Track—ins.	61.9	63.01	63.6	63.5
Rear Track—ins.	61.1	63.09	64.0	63.6
Test Weight—lbs.	4425	5010	4950	5110
Fuel Capacity—gals.	25	22.5	26	26
Oil Capacity—qts.	5 (1)	4 (1)	4 (1)	5 (1)
Luggage—cu. ft.	(16.5 with space saver spare) 14.3	13.9	14.7	17.0
PERFORMANCE				
Acceleration				
0-30 mph	3.0	3.5	3.7	3.8
0-60 mph	7.7	9.0	9.6	10.6
0-75	11.2	13.7	14.2	15.7
Standing Start ¼-mile				
Mph	88	85	82	80
Elapsed time	15.8	17.4	17.2	17.7
Passing speeds				
40-60 mph	4.2	4.5	5.0	5.8
50-70 mph	4.8	6.0	5.9	6.8
Speeds in gears*				
1st . . . mph @ rpm	N/A	48 @ 4000	45 @ 4000	48 @ 4000
2nd . . . mph @ rpm		82 @ 4000	74 @ 4000	83 @ 4000
3rd . . . mph @ rpm		89 @ 3000	84 @ 3000	90 @ 3000
4th . . . mph @ rpm				
Mph per 1000 rpm (in top gear)	N/A	29.6	28.0	30.0
Stopping distances				
From 30 mph	34' 11"	33' 11"	31'	30' 10"
From 60 mph	165'	189' 9"	160'	155' 9"
Gas mileage range	10.0-11.1	10.2-10.9	10.3-11.2	10.1-10.6
Speedometer error				
Car speedometer	30 45 50 60 70 80	30 45 50 60 70 80	30 45 50 60 70 80	30 45 50 60 70 80
True speedometer	31 45 51 61 71 81	30 45 50 60 71 80	30 45 50 59 69 79	30 44 49 58 68 78

*Speeds in gears are at shift points (limited by the length of track) and do not represent maximum speeds.

an electric shish kebab than a mere lighter. If you threw it out the window in Belfast, you'd clear the streets in three seconds.

Upholstery varies from white leather in the Thunderbird and the silken smooth nylon of the Toronado to the Grand Prix's red plush scivvie cloth, the likes of which haven't been seen since Mrs. McCabe closed her Virginia City saloon. It is the brightest red in the industry.

Grand Prix sports the strongest power train of the lot. Highway response is quick and strong. Coupled with the nimble handling, the balanced package of power and control make driving almost a thrill again.

Thunderbird's 460 engine puts all of the muscle down at the low rpm range. We actually had to walk the 'Bird out of the hole at the Orange County drag strip to keep from spinning the rear wheels. Concentration of all that torque at the low end overcomes the ravages of smog equipment to give firm response with plenty left to run all the power goodies.

Riviera ran a close third in the power arena, with Toronado bringing up a leisurely fourth. Toronado feels very sluggish under full power even though the numbers are quite respectable.

Trunk space for those long weekends away from it all are quite adequate in the Toronado and barely adequate in the other three. Riviera sports massive seals on the upstream side of the trunk opening, indicating that a previous leakage problem has been remedied.

So much for the compliments, now for the complaints. Thunderbird made up for its overkill in the cigarette lighter by the inadequacy of the driver-side ash tray. The snubber is too small. Extinguishing a cigarette at night, at speed is a major operation and it shouldn't be. This can and should be easily corrected. Wind noise in the Thunderbird is very slight, but more than we have come to expect from Ford.

The glove box is very small, however this is a side effect of leaving sufficient panel space on the right side for the dreaded air bags if, and when, they become mandatory.

Grand Prix suffers from ashtray problems as well. The little plastic cover doesn't close properly and is difficult to open once it is closed. Not enough clearance on the hinge side.

There is insufficient room between the driver seat and the door. Reaching for the left side seat belt is a squeeze.

Our GP had a sun roof, which is a great option for fresh air, but not much for head room. I'm only 5-feet-10, but I brushed the roof even with the seat fully down.

To compensate for these ills, GP does have a superb steering wheel. If that compensates. Just the right size and spongy enough for a solid grip. It is also adjustable, as are all in this class.

Riviera has fewer difficulties. Partially open windows create excessive noise and there is a fierce breeze passing through the ventilation system at the lowest setting.

The design of the roofline inherently limits the vertical dimension of the back window, leaving rearward visibility only satisfactory.

The only two problems we noticed with the Toronado was the very slow recovery rate of the suspension on large bumps, giving a roller-coaster effect for a couple of cycles, and the impression of size imparted by the front hood design. It is a bit boxier than the other three, leaving the driver well aware of fact that he best not blunder carelessly into tight places. It was a trifle intimidating.

All four vehicles do the job of imparting an image of prestige and attainment, with a small handling bonus for the driver. Toronado and Thunderbird do the best job. Thunderbird, the original, gets the edge. Riviera comes on less stately, but with distinctive styling. Unfortunately, the torpedo back is to pass into history with the 1974 model. Grand Prix is the least ostentatious, but the most fun to drive.

Price may help you with the final decision. Thunderbird is $8105, Toronado is $7862, Grand Prix is $6716 and the Riviera goes for $6516. Prices are without transportation charges and loaded. ■

LATEST GENERATION – The 1972 Oldsmobile Toronado Custom is the latest generation of this famous marque, and continues the tradition of its predecessors.

ROAD TEST

THE PERSONAL LUXURY CARS

Thunderbird, Riviera, Toronado: Big, quiet, expensive.
Monte Carlo, Grand Prix, Cougar, Cassini Matador and the new Torino Elite:
Three new entries and veteran take after the leader a in the field.

BY JIM BROKAW

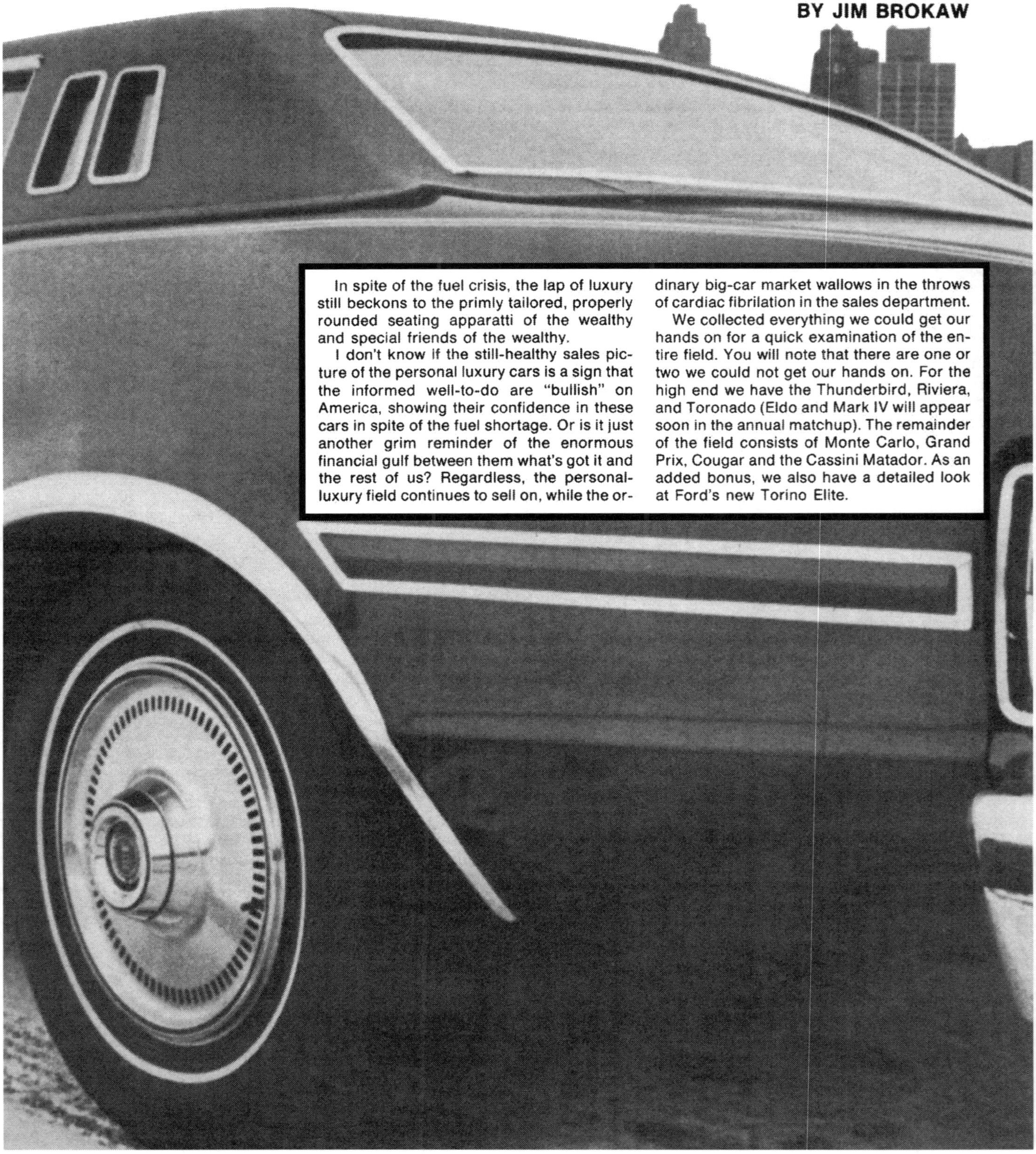

In spite of the fuel crisis, the lap of luxury still beckons to the primly tailored, properly rounded seating apparatti of the wealthy and special friends of the wealthy.

I don't know if the still-healthy sales picture of the personal luxury cars is a sign that the informed well-to-do are "bullish" on America, showing their confidence in these cars in spite of the fuel shortage. Or is it just another grim reminder of the enormous financial gulf between them what's got it and the rest of us? Regardless, the personal-luxury field continues to sell on, while the ordinary big-car market wallows in the throws of cardiac fibrilation in the sales department.

We collected everything we could get our hands on for a quick examination of the entire field. You will note that there are one or two we could not get our hands on. For the high end we have the Thunderbird, Riviera, and Toronado (Eldo and Mark IV will appear soon in the annual matchup). The remainder of the field consists of Monte Carlo, Grand Prix, Cougar and the Cassini Matador. As an added bonus, we also have a detailed look at Ford's new Torino Elite.

Taking it from the top, the most distinguishing characteristic of the T-Bird, Riviera and Toronado is *size*. They look big and they are big. Perhaps a bit too big for their designed function. For example, Thunderbird, the vanguard of the personal luxury market, with its four-door model in 1958, has grown from a wheelbase of 113 inches to 120.4 inches and added to its overall length from 205.4 inches to 224.8 inches. The leg room, however, which we are sure was measured by different standards back then, has decreased from 43.4 inches to 42 inches. To Ford's credit, head room has increased by three inches. Buick and Olds have little reason to snicker, as the Riv sits on a 122-inch wheelbase riding over 18 feet long, while Toro matches the 122-inch wheelbase but stretches to a full 19 feet in length. Is it really necessary to have that much machine to achieve the desired smooth ride and air of elegance required for a *personal* luxury vehicle?

So much for size. On the plus side, the magnificent three do have a smooth ride and a first-class interior. Wide-ribbed, crushed-velour upholstery in horizontal and square tucks. Deep, cutpile carpeting. Wood grain dashes that look like real wood. Optional automatic air conditioning systems. Optional stereo everything. Quiet—very, very quiet. Much effort has gone into the science of reducing wind and road noise in all three. Ford is the pioneer in the silencing business, but its edge in this field is very slight.

Both Ford and GM have clustered instrument panels directly in front of the driver, which is very good. All of the gauges and knobs are close at hand. Buick made a boo-boo by placing the transmission selector so that it interferes with access to the radio knobs when the trans is in drive.

All three had six-way power seats on the driver's side. In addition to good seating comfort, the six-way can be rigged to taste. Naturally, it is an optional extra.

Riviera has followed Ford's lead and installed a remote controlled right-hand mirror. This is a necessary option. Rearward visibility is good in all three, but the addition of the controllable right side mirror gives the driver full coverage. With cars of these dimensions, you need it.

There is very little edge to be enjoyed in the realm of comfort and luxury by any of the three. They all share basic features, and all are well executed. It's really a matter of product loyalty and styling taste. In the area of power, we found a bit of a difference. Thunderbird's 460-4V did not exhibit the usual strong torque at low rpm. It did show an unusual thirst for fuel, logging an unimpressive 11.69 miles per gallon over our 73-mile test loop.

Toronado showed neither strength nor weakness in the area of engine response. Power from the 455-4V was adequate. Fuel economy was a near-respectable 13.36 miles per gallon.

Riviera's 455, which is not the same engine, having a different bore and stroke and a different set of engineers doing the development, showed the most strength, being quite positive in the low range. It was equally positive for its class, in fuel consumption, carding a 14.04 mpg. It would seem, at least for the heavyweights, that having sufficient power contributes to good fuel usage.

Although Toronado is front-wheel drive, in normal usage there is no detectable advantage save the absence of the hump from the trans tunnel.

None of the three were easy to park or particularly comfortable on city streets. All were more than comfortable on the highways.

Moving down a class to the more affordable personal luxury models, we have a bit more distinction and individuality. The surprise of the group was the Cassini Matador. We felt a bit presumptive pitting the Matador against the Cougar, Monte Carlo and Grand Prix, since the Matador is really just a two-

MOTOR TREND SPECIFICATION DATA

SPECIFICATIONS	THUNDERBIRD	RIVIERA	TORONADO
Displacement—cu. in.	460	455	455
HP @ RPM	220 @ 4000	230 @ 3800	230 @ 3800
Torque: lbs.-ft. @ rpm	355 @ 2600	355 @ 2200	370 @ 2800
Compression Ratio	8.0:1	8.5:1	8.5:1
Carburetion	4V	4V	4V
Final Drive Ratio	3.00:1	2.93:1	2.73:1
Width—in.	79.7	80.0	79.5
Wheelbase—in.	120.4	122.0	122.0
Overall length—in.	224.8	226.4	228.0
Height—in.	53.0	53.7	53.3
Curb Weight—lbs.	5270	4930	4998
Fuel Capacity—gals.	26.5	26.0	26.0
Storage Capacity—cu. ft.	13.4	N/A	17.0
Base Price	$6542	$6032	$5759
Price as tested	$8607	$7358	$7671

COUGAR **16.2** mpg
CASSINI MATADOR **14.3** mpg
GRAND PRIX **14.3** mpg
MONTE CARLO **14.0** mpg
RIVIERA **14.0** mpg
TORONADO **13.3** mpg
THUNDERBIRD **11.6** mpg

Fuel consumption calculated on a 73 mile loop of city, suburban, freeway and hilly roads. Speeds did not exceed 60 mph.

door sedan with a gussied-up interior and the other three are specialty type cars, but the *What's-a-Matador* more than held its own.

All the staff members who drove the Matador agree that it has the best combination of ride and handling. It is nimble and smooth in all street situations. The 401-4V engine is strong in all normal driving modes and clocked a very respectable—for its class—14.31 miles per gallon. Matador's styling is quite unique, being specifically designed to go fast on the NASCAR oval tracks. The first Cassini Matador we saw was black with a gold vinyl roof. It exuded class and luxury. The one we tested was white with a goldish-tan vinyl roof. Get the black one.

COUGAR	GRAND PRIX	CASSINI MATADOR
400	400	401
168 @ 3800	225 @ 4000	235 @ 4600
310 @ 2000	330 @ 2800	335 @ 3200
8.0:1	8.0:1	8.25:1
2V	4V	4V
2.73:1	2.93:1	3.15:1
78.6	77.9	77.44
114.0	116.0	114.0
215.5	217.5	209.35
52.4	52.7	51.76
4510	4400	4056
22.0	25.0	24.5
16.5	N/A	17.3
$4301	$4727	$3149
$5059	$6495	$5257

Oleg Cassini, the jet-set clothes creator, has designed an interior combination of deep black nylon knit cloth, copper medallion buttons and copper dash and steering wheel insets that are pleasing to both the eye and taste. In spite of Matador's plush interior and very good handling, there are a few short points. The distance from floor pan to the top of the seat cushion is high enough to rob the driver of head room. Fortunately, our test car had a reclining seat back. By tilting a few degrees, we were able to get head clearance. Unfortunately, it was the standard AMC recliner with the worse control device since Pontiac's lumbar lump. There is no convenient way to operate the AMC seat handle. American Motors would do well to develop a more functional control lever. Like the one in the Grand Prix.

Ten thousand cheers for Pontiac. The Grand Prix has an infinitely adjustable seat back with an easy-to-use handle. And it works. There is no reasonable excuse why every domestic vehicle, regardless of price or class, can't offer a similar option.

The Grand Prix tended to leave our staff with a very positive impression. Either they liked it very much, or they disliked it very much. No neutrality.

Ride and handling in the Grand Prix were a satisfactory combination—not as good as the Matador, but better than the three high-buck models. The 400-4V engine was adequate, not exhibiting any shortcomings, nor any signs of greatness. Mileage matched the Matador with an identical 14.31 mpg. Certainly respectable.

Grand Prix's dash is the GM cluster with a large clock instead of a tach, on what may be the wrong side of the panel, since we were continually noting the time instead of speed while making it down the Santa Ana freeway. It is easier to see the clock.

One consistent complaint by all was the shape and location of the door pull slots in the arm rest. They are in an awkward location—too far back—and coupled with the length and weight of the door, it makes for a bit of contortion to get the door fully closed on the first try with all the windows closed.

The front buckets, which are very wide and padded, sculpted and contoured, were quite comfortable to me, but not to others. Dealer's choice.

Cold starting characteristics of the 400 engine were somewhat balky. Once through the warm-up stage, it performed very smoothly.

The Grand Prix is a very individualized package. Everything on it is specific in its intent. There are no compromises. It is a strong statement of personality. If it fits, you'll love it. If it doesn't, well, that's your choice.

Cougar is not the Cougar of yesterday and very definitely not the cat of several years ago. It is a Montego dressed out to look like a mini Mark IV. It is neither mini, nor is it a Mark IV. It is a very good balance of ride and handling with the edge toward handling. On the street it is very stable and corners quite flat, with the resultant penalty in slight harshness.

The interior is nicely laid out, as are all the cars in this class, with a full set of engine monitor instruments. We had the relatively plain Cougar with a vinyl interior and no power comfort options. I would have preferred power windows, and for the Cougar, I think you would too. It comes off as adequate for a personal luxury car with the base interior, but it talks much louder with the brougham set up.

We had a 400-2V under the hood that's quite satisfactory for all street use, but it lacks inspiration. It more than compensates with a noteworthy 16.22 mpg on the fuel test loop. With gas at a premium, maybe you can live with the lack of suds.

Some felt that the tachometer was a bit much for a personal luxury car. Not anymore. If you are really playing the featherfoot and squeezing every drop of fuel out of your tank, drive by the tach. Smooth and easy means more miles.

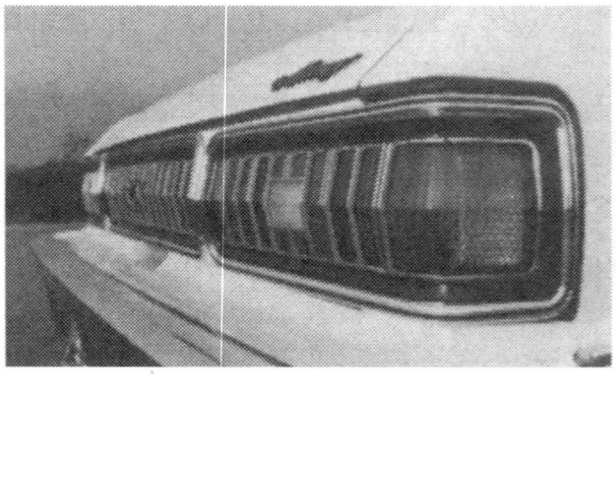

The leader of the pack is Monte Carlo. Ride and handling balance is very close to the Matador. Dash and interior appointments are of quality material and well laid out. Swivel bucket seats are still a bit awkward, but this is more a fault of habit than a fault of the seat. Once you adjust to round-and-round instead of tilt-and-slide, they are fine. Best lower back support of the lot. One drawback. In order to bring the doorside seat belt forward, you either have to reach over the edge of the seatback, which takes a bit of doing, or open the door. There is minimal clearance between the seat

TORINO ELITE

We chose to break out the Torino Elite, because unlike the other cars, it is not a production model, and it was tested in Detroit around a snow storm. We did not run it on our standard fuel loop.

It is very definitely Ford's answer to Monte Carlo's 300,000 plus sales in 1973. Ford did not have anything in their model lineup to go head to head with Monte Carlo. Thirsting after a piece of all that action, Ford targeted Cougar to compete in the high end of the corporate line and went to the styling studios for the Ford division entry.

In spite of the hurried job, the end product shows no signs of the traditional penalty for haste. The GTE, Gran Torino Elite, is based on the Montego frame, which is stronger and beefier in the front end than Torino's. Front and rear spring rates are higher than Torino, shock valving is different and there is a larger front sway bar. The rear sway bar, optional on Torino, is standard on the Elite.

Basic sheetmetal is Montego, with Cougar rear fenders, but front hood and grille are unique to the Elite. Wide chrome wheel lip moldings and T-Bird side rub strip give it the top of the line look Ford is hoping for.

The sound deadening package is the best available on an intermediate, exceeded only by the package in the T-Bird and Mark IV. It also has 25 ounce cutpile carpeting that runs up over the kick panel on the door.

The standard Torino H78-14 tires on a 5½ inch wheel have been upgraded to H78-*15* on a six-inch wheel. The optional handling package includes HR70-15 tires on 6½-inch wheels.

Although the power steering is the same as Torino, the heavier suspension and larger wheels and tires give better "feel" than the standard arrangement.

On paper, Ford has created a formidable opponent for the Monte Carlo. You'll be able to judge for yourself when they hit the showrooms this month. ■

SPECIFICATIONS	MONTE CARLO	TORINO ELITE
Displacement—cu. in.	400	400
HP @ RPM	180 @ 3800	168 @ 3800
Torque: lbs.-ft. @ rpm	290 @ 2400	310 @ 2000
Compression Ratio	8.5:1	8.0:1
Carburetion	4V	2V
Final Drive Ratio	2.73	
Width-in.	77.6	78.6
Wheelbase-in.	116.0	114.0
Overall length-in.	213.1	216.1
Height-in.	52.7	52.8
Curb Weight-lbs.	4215	4291
Fuel Capacity-gals.	22	26.5
Storage Capacity-cu. ft.	14.7	16.5
Base Price	$3780	TBA
Price as tested	$5861	TBA

The 400-4V engine logged 14.04 mpg, which is right in the ball park for the personal luxuries, but lacked character on the street. Adequate, but uninspiring. I'm afraid that in the search for better fuel economy, inspiration will be a rare quality.

We learned that finding and fixing seat belts to be universally difficult on all the cars tested. We also found that cold engine starts are spotty. Sometimes just a little rough running, sometimes it took several tries. This is the fault of the smog adjustments.

All the mid-line personal luxury cars had a good balance of ride and handling with the Matador being the best.

All had good instrumentation, with the Grand Prix layout being the best in spite of the clock.

Rear seat ingress and egress was universally awkward with the hanging shoulder harness.

A couple of cars had canted steering columns, likely a concession to all the underhood hardware. The Matador's was a bit extreme.

None of them had disqualifying faults. A lot of work has gone into suspensions and seating for the driver, and you the customer are the beneficiary. Let us hope that the engineers keep at it.

OLDSMOBILE TORONADO

SPECIFICATIONS

ENGINE

Type	OHV V-8
Displacement, cu in	455
Displacement, cc	7453
Bore x stroke, in	4.13 x 4.25
Bore x stroke, mm	104.8 x 108.0
Compression ratio	8.5:1
Hp at rpm, net	215@3600
Torque at rpm, lb/ft, net	370@2400
Carburetion	1 4-V

DRIVELINE

Transmission	3-spd auto
Gear ratios:	
1st	2.48:1
2nd	1.48:1
3rd	1.00:1
Final drive ratio	2.73:1
Driving wheels	front

GENERAL

Wheelbase, ins	122.0
Overall length, ins	227.6
Width, ins	79.7
Height, ins	53.3
Front track, ins	63.6
Rear track, ins	63.5
Trunk capacity, cu ft	17.0
Curb weight, lbs	5040
Distribution, % front/rear	58/42
Power-to-weight ratio, lbs/hp	23.4

BODY AND CHASSIS

Body/frame construction	separate
Brakes, front/rear	vented disc/drum
Swept area, sq in	350.8
Swept area, sq in/1000 lb	69.6
Steering	recirc. ball
Ratio	20.4:1
Turns, lock-to-lock	3.2
Turning circle, ft	45.6

Front suspension: Independent, upper and lower control arms, torsion bars, tubular shocks, anti-roll bar

Rear suspension: Beam axle, four link control arms, coil springs, tubular shocks

WHEELS AND TIRES

Wheels	15 x 6.0
Tires	JR 78-15 General Steel Radial
Reserve load, front/rear, lb	241/1139

INSTRUMENTATION

Instruments: 0-100 mph speedo, trip odo, fuel level, vacuum gauge

Warning lights: directionals, high beams, gen, oil, brake, lights, engine temp, coolant level, low fuel, external lamp, seat belts

PRICE

Factory list, as tested: $9753.55

Options included in price: Air cond–$37; 6-way seat–$126; tinted glass–$64; floor mats–$43; door edge guards–$8; litter contain–$6; cruise control–$79; tilt wheel–$53; trip odo–$19; AM/FM stereo tape–$341; auto antenna–$40; indicator pkg–$24; appearance opt–$281; power door lock–$90; power trunk release–$17; side molding–$29; rear defog–$78; remote mirror–$28; illum visor mirror–$40; wsw tires–$60; cornering lamps–$41; lamp monitor–$43; economy meter–$25; rear bumper–$12; convenience grp–$29; cooling system–$21; Cal emiss–$50; dest. chg–$421

TEST RESULTS

ACCELERATION, SEC.

0-30 mph	4.0
0-40 mph	5.8
0-50 mph	7.9
0-60 mph	10.5
0-70 mph	14.0
0-80 mph	18.6
Standing start, ¼ mile	17.6
Speed at end ¼ mile, mph	78.3
Avg accel over ¼ mile, g	0.20

SPEEDS IN GEARS, MPH

1st (4000 rpm)	48
2nd (3800 rpm)	76
3rd (3800 rpm) (calc.)	111
Engine revs at 70 mph	2400

SPEEDOMETER ERROR

Indicated speed	True speed
40 mph	39 mph
50 mph	49 mph
60 mph	59 mph
70 mph	69 mph
80 mph	79 mph

INTERIOR NOISE, dBA

Idle	50
Max 1st gear	69
Steady 40 mph	59
50 mph	62
60 mph	64
70 mph	67

BRAKES

Min stopping distance from 60 mph, ft	196
Avg deceleration rate, g	0.61

HANDLING

Max speed on 100-ft rad, mph	30.1
Lateral acceleration, g	0.61
Transient response, avg spd, mph	21.9

FUEL ECONOMY

Overall avg, RT cycle	15.5 mpg
Range on 26.0 gal tank	403 miles
Fuel required	unleaded

RATING

Graph Of Recorded Data Expressed in Percentage of 100 (100 = best possible rating)*

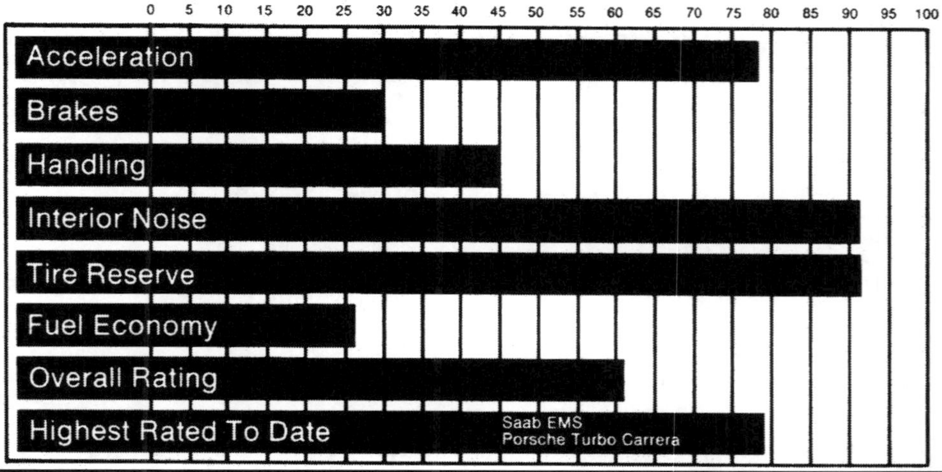

*Acceleration (0–60 mph): 0% = 34.0 secs., 100% = 4.0 secs.; Brakes (60–0) mph: 0% = 220.0 ft., 100% = 140.0 ft.; Handling: skidpad lateral accel., 0% = 0.3 g, 100% = 0.9 g, transient response, 0% = 20 mph, 100% = 25 mph (average skid pad and transient response for overall handling percentage); Interior Noise (70 mph): 0% = 90.0 dBA, 100% = 65.0 dBA; Tire Reserve (with passengers): 0% = 0.0 lbs., 100% = 1500 lbs. or more; Fuel Economy: 0% = 5 mpg, 100% = 45 mpg or more. Test Equipment Used: Testron Fifth Wheel and Pulse Totalizer, Lamar Data Recording System, Esterline-Angus Recorder, Sun Tachometer, EDL Pocket-Probe Pyrometer, General Radio Sound Level Meter.

Oldsmobile Toronado

No brontosaurus ever went into the peat bog with greater dignity.

Don't be fooled by the dinosaur. Unlike the huge reptiles which ruled the world for eons, the Oldsmobile Toronado will not crash one day into oblivion, leaving only scattered bones and a memory of giants. Instead, it will fade into the folklore of the automotive world, achieving a kind of immortality.

There seems to be little doubt, though, that if this year isn't the final year of production for the gargantuan Olds, then it must be the next or the next after that. They may shrink it a bit and hang the name on some other body style, but unless it weighs the two-and-a-half tons of the current model . . . well, you know. It just won't seem like a real Toronado.

Ever since the car's inception (in 1966) it has gotten bigger and heavier and plusher. In the last decade it gained 500 lbs., 16 inches of length and opera windows. It has faithfully followed the trend to more vertical styling, and it stands now as a remarkably clean design for such a huge car, epitomizing the Detroit Dream: longer, lower, wider.

The original 425 cid V-8 has grown to 455 cubic inches, and performance has stayed about the same; in 1966 you could expect your Toronado to do 0–60 in about 10 seconds, and your '76 car will do it in 10.5, with a top speed within one mph of the older car. Fuel economy stays in the same league as well (but only after concerted efforts last year by the Olds engineers), with 15.5 mpg being the best we recorded in our cycle, as opposed to the older Toronado's 13 mpg.

There is a great difference in what you pay today to have that 227 inches of Moby Olds in your garage, though; in 1966, it would set you back about six grand, but today you're looking at almost ten thousand bucks for an equivalent package . . . although it should be noted that today's car has many gimmicks not available ten years ago.

That the Toronado should have found such a comfortably uncontroversial nesting place in the American consumer's arms would surprise few today, but ten years ago a front-wheel-drive car—especially an *American* fwd car—was a major automotive event in itself. Still, after driving the latest Toronado, it's easy to see why it causes little controversy; aside from its nice, flat floor (where the driveline bulge of a conventional car would be) and occasional chirp from the front tires under full throttle, it could easily be any other big American car. The virtues of the fwd design—better space utilization, better traction in foul weather and fewer driveline bits—all seem to take second place to the more traditional features of this kind of car, features like plush interiors, soft rides and slick styling.

Indeed, so limited is the interior space (longitudinally, not laterally) of the Olds that you can't help but wonder why they bothered. The rear seat room is worse than that of last month's biggest test car, the Buick Riviera, which was no ergonomic miracle itself. Unless the person driving likes to have his knees jammed under the dash, the guy behind him will have *his* knees jammed into the seatback . . . which hardly seems useful in a car almost 20 feet long.

However, it seems utility is not high on the Toronado owner's priority list, because month in and month out, Olds sells about 1500 of them. This in itself lends a kind of exclusivity to the car which may help to explain its continuing popularity. The numbers aren't big (especially compared to the Toronado's kissin' cousin, the Eldo, which sells double that figure), but they're *steady*, much like the traditional Olds buyer himself. Chances are that people in search of a car without too much ostentation, too much aggressiveness, too much visibility, or maybe even too much anything, are drawn to the Toronado like bees to flowers. In that way, the Olds people have managed to keep the Toronado in touch with an identity tangible enough to make you take things a little easier from the minute you begin to drive the car. Which is an admirable achievement, considering the fact that they had to do it with the same parts bin as everyone else. Like the Eldo, the Big T has a real personality, which for some people is far more important than all the shoulder-room and rear seat knee-room figures ever put together.

In its own way, the Toronado is as stately as a Rolls; while not being capable of feats of hard driving, it seems to produce an atmosphere of 19th Century timescales in which hard driving figures not at all. If you don't look too closely at the actual materials (which are likely to turn out to be cleverly printed halftone for the wood trim and injection-molded plastic for the heavy chrome trim fittings) you could even feel yourself in some 19th Century time warp, surrounded by cool elegance and soft music.

The key is not to look too close, because once you do, you can't stop . . . and your ten grand gilded carriage is apt to turn into a pumpkin fast. Cars like the Toronado should be appreciated from a distance (if at all) with careful suspension of disbelief; the minute you begin to take them seriously, they conjure endless visions of gulped fuel and lube jobs and cracking vinyl tops and electrical goodies gone on the blink.

Likewise, you must not expect the car to *perform* in the contemporary sense; its 2.5 tons of weight galloping down the freeway can be halted in large areas of real estate—196 feet from 60 mph was our best stop—and will wallow around uncomfortably on hard corners (0.61g on our skidpad). It encourages you to drive sedately, and for maximum enjoyment—or maybe any enjoyment—you'd better listen up and ignore those pesky sportscars buzzing around your windows.

The decision made back in the '60s by Olds to build a fwd car was, for the times, a radical one with a thousand potential pitfalls. The car built for those times had enough technical goodies to keep the owner interested in his car for quite awhile. They even raced the thing (remember Pike's Peak?) to demonstrate that it wouldn't fall off cliffs because of the fwd regardless of what the boys in Elmo's Super Service Station downtown muttered. From that heritage it is a little disappointing to look at the newest Toronado and see that it all seems to have been for nothing; it has been diluted and blended in with the mainstream cars for so long that most people probably don't even know that there is anything different about it at all. And that's a shame, because—unlike the dinosaurs—it means that when Olds finally lays the car to rest in some Detroit peat bog, no one will notice. ■

Toronado -a prophetic dream machine

The Oldsmobile Toronado remains, with the Cadillac Eldorado, the only front-wheel-drive car that the giant General Motors Corporation produce. And with their professional interest in front-wheel drive for the coming new generation of GM lighter-weight passenger cars, the Toronado, after 10 years of production, assumes a new significance

By Chris Goffey

IN 1955, General Motors produced one of their styling exercise dream cars, called the La Salle II. It differed from many of the dream car shells of the time, since GM Engineering had co-operated with Styling to fit the car with a power unit. The car was displayed at the General Motors "Motorama" show in that year and attracted great attention from the pundits, not simply for its futuristic looks, but through the use of front-wheel drive. The GM Power and Transmission Development Groups first became involved with what was known as "UPPs" (unitized power packages) and front-wheel-drive designs the year before at General Motors, and the La Salle was the first public display of the way their thoughts were developing.

The design for the La Salle considered the use of an automatic transmission, but time limitations did not allow this avenue to be fully explored. From this early look at the fwd concept came a number of ideas, some of which finished up being put into practice in 1962 for Oldsmobile bodies by General Motors Engineering Staff. However, the separate Oldsmobile Division engineers were also interested in the front-wheel-drive packages displayed in the La Salle and they started their own similar and parallel studies into front-wheel-drive designs. They, too, used their ideas in Oldsmobile F85 bodies, and it was at this point that Central Engineering staff and Oldsmobile engineers came together on the development of a practical and feasible design. The code name. . . . XP-748.

XP-748 turned out to be the Oldsmobile Toronado, unveiled in the autumn of 1965, a dream car in production which created enormous interest on both sides of the Atlantic.

The Toronado was very much a General Motors — as opposed to Oldsmobile — product. Once they had settled on the idea of a front-wheel-drive concept and had decided — after a lot of wrangling and experimenting with smaller cars of the F-85 and Opel size — that the product they wanted would be large, luxurious and an Oldsmobile, the responsibilities for development were allocated.

Oldsmobile were to look after development of the engine, front and rear suspension and sub-frames; the Fisher Body Division were to assume responsibility for the body structure and the passenger accommodation; Hydramatic Division were to look after the transmission; Buick would develop the differential; and Saginaw Steering Gear Division would look after development of the front-wheel-drive units.

Oldsmobile had already done a great deal of work on front-wheel drive before this stage, indeed there was some suggestion that the F-85 might at one stage go into production as a front-wheel-drive vehicle. They had experimented with the front-wheel-drive set-up in a variety of conditions and had come to the conclusion that it was well worth employing for straight-line stability, cornering and general performance in wet and winter weather.

There were, however, important provisos. The car had to use existing power plants, three-speed automatic gearbox and differential components. The project took 10 years to get straight, but gradually, a picture began to emerge.

The power plant was to be the 429 V8 production unit, at that time the largest GM passenger car engine available. It gave maximum performance in the minimum weight package and, for the first time in many years, the designers had the freedom to abandon the four-degree rearwards inclination to reduce prop shaft angularity and could tilt the engine forwards if they so desired. Original designs called for the crankshaft of the engine to be extended to provide the input to a torque converter, but GM were not happy with the idea of some engines from the engine plant having one type of crank and with the problems of separating the transmission from the engine.

Subsequent re-designs eliminated the need for the extended crank nose. The only other modification for the unit was the use of a new intake manifold that had minimal height and was made of alloy to reduce weight. A low pancake air cleaner was also designed for the unit.

The final design for the transmission was chosen after a variety of layouts had been built and assessed. The transmission was placed on the right of the engine (looking from the driver's seat) with the torque converter on the end of the engine as normal.

The fluid flywheel gave torque multiplication, flywheel inertia and a fluid cushioning effect to smooth out engine torque fluctuations. Then a multiple Morse chain took the drive from the torque-converter turbine to the almost-standard, three-speed, Hydramatic gearbox unit.

Given that decision, the two most important problems to tackle were chain noise and durability. Hydra-Matic and Oldsmobile had been looking at both these factors, and dynamometer chain test installations were built to study high-speed use, and a wide range of loads and speeds.

Hydra-Matic faced a problem in developing a basically standard automatic that revolved in the opposite direction to normal. Gear helix angles were thus reversed to control thrust direction and free-wheeler and band units were revised, but basically the components used were standard.

Saginaw, in the meantime, had been working away on inboard constant velocity joints which would have sufficient capacity for such a powerful engine, within the confined space of the engine bay. They came up with a six-ball joint in combination with a ball spline for axial travel allowance.

A great deal of work was still going on to contain the noise of the chain drive. Modified tooth profiles, slack adjusters to control position and tension, and isolators to prevent noise transmission were all investigated on a dynamometer.

In the end, seven units were built; three were kept by GM Engineering Staff and four went to Oldsmobile.

With the power plant in their hands, after 10 years of development, Oldsmobile were ready to build the car they hoped would shake America.

The Structure and Suspension engineers had already decided what they wanted to see in the new car. Front-wheel drive allowed them to build an independent front sub-frame to carry engine and independent front suspension. The sub-frame would be mounted to the body through insulators to cut down vibration and road surface noise and they saw the possibility of using a simple rear suspension and a low flat floor in the body.

The picture sequence shows from left, 1969 facelift; further modifications for 1970; and three variations on the fundamental shape change from 1971, through 1972 to 1973. The current model, due to be phased out soon, follows this line

Toronado's original distinctive lines have become bulkier and more commonplace over the years, reflecting the use of the same basic body as the Buick Riviera and other GM "personal" cars.

GM had considerable experience of torsion bar suspension from the early 1930s but the suspension engineers wanted to try a new bar which they had developed a couple of years before the Toronado concept. It only needed to be about one-half of the length of a conventional torsion bar. This meant that the rear end of the bar could be attached to the engine sub-frame rather than to the bulkhead. At the rear, a simple beam axle and Hotchkiss single-leaf springs were used, GM having had a great deal of experience with such springs from the Chevy II design of 1962.

In the summer of 1962 General Motors Styling staff were asked by Oldsmobile to design a striking two-door body to clothe the new engineering packages. It was to be a car in keeping with Oldsmobile's character and engineering quality and the design brief was that it had to have "functional" lines. The car had also, according to Oldsmobile, to look "sporty", with a long bonnet and a chopped-off boot.

Styling's Advanced Design Studio No. 3 was allocated to the project; the door locks and security moved in and the design evolved under the supervision of Oldsmobile designers, but isolated from other production design activities.

According to Bill Mitchell, recently retired head of GM Styling, many advanced ideas look good on paper, but the real test comes when it emerges in solid, three-dimensional form. Thus, a full-size clay model of the Toronado emerged, with frequent reviews from styling management' who evaluated every aspect. The first showing of the clay full-size model was in February 1963 and took place outdoors, for optimum light conditions. First reactions were enthusiastic — and approval was given to the basic design.

In the course of evolution, the waistline was lowered and the rear boot line lengthened, to improve the profile. Wheel diameter was increased to give more emphasis to the wheels. The first glass-fibre model, an exterior only version, was ready for review in July 1963 and an updated model with full interior and all hardware fittings was ready in the early summer of 1964.

Production of the Toronado on a volume basis began in August 1965 marking the culmination of 19 months of intensive preparation involving assembly planning, testing of facilities, systems, and methods; and the training of personnel. The Toronado production was planned to be accomplished with facilities separate from those used in the production of other Oldsmobile car models and completely new lines had to be laid down.

Testing of the new car produced some alarms early on. The first test cars were 1964 Oldsmobile 88 bodies converted to flat floor, with the Toronado separate sub-frames supporting the front suspension and the drive train ending under the front seat with two body mounts. Four compression body mountings were used to join the frame to the body at the facia. However, out at the Desert Proving Ground, well away from prying eyes, the first test cars demonstrated a lack of torsional and beam strength on rough road surfaces and suffered resonant shake on freeways and excessive road noise. A development car was reworked in an around-the-clock operation to cut and weld-in structural channels extending the sub frame backward. It worked and design changes were rapidly processed to meet the lead times.

The new stronger body frame reduced shake and permitted the use of softer body mountings that suppressed road noise and vibrations.

The tyre manufacturers were brought into the game to design an 8.85 x 14 low-profile tyre of conventional construction with stiffer sidewall construction to meet the needs of front-wheel drive. Final design was a special T-FD (Toronado Front Drive) manufactured by Fires tone with a special high-grip tread.

Other problems started to show at the prototype stage. The Toronado was to have power steering as standard equipment and early experience showed that steering linkage rigidity was very important in order that precise control of the front wheels should be maintained during acceleration, braking and cornering, when the enormous torque of the big V8 imposed great loads on the suspension and steering. A variable ratio was chosen with a low centre ratio to give quick response, linked to higher full-turn ratios for reducing the load pressure when parking.

Ride was the next area to be investigated at this stage. Concentrated ride and handling programmes were conducted on the cars in Michigan and Arizona, since Oldsmobile had made one very

Drawing of original chassis shows that, engine / transmission package aside, it is a fairly conventional American car. The front of the chassis with power unit and running gear installed is used in some luxury motor-homes; chassis are taken off the production line and cut in half for the purpose

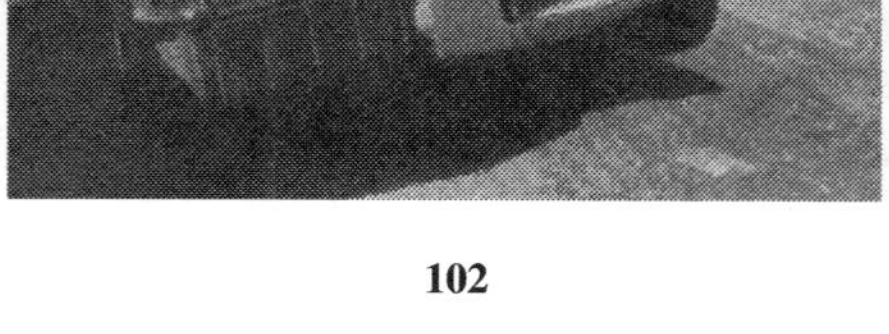

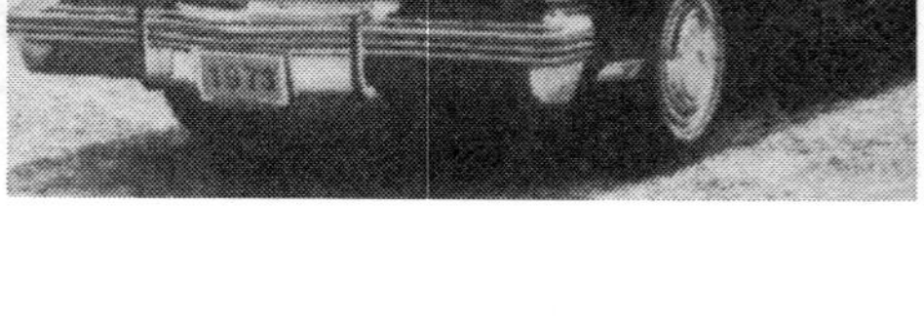

important stipulation early on in their decisions for the new "American Grand Touring Car" as it came to be known. No compromise was to be made on car handling. Oldsmobile wanted the Toronado to astonish Americans with its ride and handling and to take "a real step forward in the area of car behaviour".

The Toronado kingpin-to-wheel-centre relationship was an example of this design philosophy; it was unique. A ½in. outboard scrub radius and 2in. wheel offset meant that scrub and wheel fight were held to a minimum. The resulting kingpin inclination of 11 degrees also gave the designers the marked directional stability and the self centring forces they required.

Brakes were an early design decision and a variety of types were tried, including a disc brake set-up. However, in the end, the engineers opted for cast iron drums and conventional linings. The wheel offset requirements meant that the wheels would have to have wrap-round, deep-set drums and there was a great deal of early concern about cooling. In the end they designed-in 10 large openings in the wheel together with finned drums and met their fade and wear criteria.

The more the testing went on, the more the car was refined and problems ironed out. A special exhaust system was devised with spherical flexible couplings to allow freedom of power train reaction during torque reversals, without having the exhaust banging the car floor.

To eliminate completely any noise during torque reversals, either starting or driving, a rubber coupling was built into the right-hand drive shaft, which allowed seven degrees of axle turn against rubber doughnuts before solid contact was made, a full 300 lb. ft. torque development.

If the engineers thought that their mechanicals were unique, the body engineers, given the job of translating the Styling Department's ideas into sheet metal, were just as enthusiastic about the car. The flat floor and the massive length of the door opening (63 inches from jamb to jamb) allowed them exceptionally good rear-seat entrance and exit conditions.

The seats received a great deal of attention. Standard fitting was, of course, the three-abreast American standard bench, but Oldsmobile must have reckoned that somebody out there would need some positive locations to use all that cornering power they had provided and so "stato" bucket seats were designed for the car as options. These came with a reclining passenger seat and adjustable head restraints, and they waxed lyrical about their through-flow ventilation system which early on elminated the need for side vents on the car, aiding the reduction of wind noise.

The Toronado was styled with retractable headlamps and given the goal that the lamps should come up quickly and reliably through 56 degrees, and be capable of operating efficiently in heavy icing conditions. The engineers discarded electrical operation, since they reasoned that it required gear ratios which would not give acceptable operating speeds. A vacuum-actuated system proved the answer and was engineered to give a mighty potential pull of 200 lb to the hinge lever. That made sure that they came up quickly. A reserve vacuum tank in the system gives three raise and lower operation with the engine off and a safety notch in the raising mechanism prevented accidental lowering through vacuum failure while on the move (Colin Chapman please note). The head-lamp system was installed in a 1965 Oldsmobile and driven all over the country in every altitiude and climatic condition the engineers could find. Only after 100,000 cycles were completed in the lab at temperatures from −20 to 250 degrees F was the system approved for production.

After a pilot building programme, when the engineers and testers once again descended on the car and put it through the rack, production started and the world was about to see the new look from Oldsmobile. It had taken seven years and one and a half million test miles to reach this point. How would the world react?

Autocar took their first look at the car in January 1966 when Geoffrey Howard road tested the Toronado. He referred to the scepticism existing among Europeans about a powerful 7-litre engine driving through the front wheels. "If one were to shut off power suddenly midway through a corner, would an understeer/oversteer transition make it almost unmanageable for ordinary mortals?" he asked.

At the time it was tested, the Toronado had the widest track of almost any current production car (5ft 3in. at the front) and even today that figure is exceeded only by the Panther De Ville.

The road test referred to the almost uncanny freedom from mechanical commotion or wind around the body, right up to the maximum speed, which was just over 125 mph. This, we said, was its outstanding virtue.

The Toronado ran very easily and accurately round corners and had a very high ultimate cornering level. However, as might be expected with a car weighing some 2¼ tons, the brakes proved inadequate for the *Autocar* fade tests at MIRA. One full-energy stop from 90-100 mph had them cooked beyond capacity long before the car had come to rest.

V8 engine and automatic transmission (left) make a very neat package. The wide, rubber-damped Morse chain has given no service problems. Cross section of power train (below) shows arrangement of transmission parts, the cost of which has prevented wider application on American cars so far

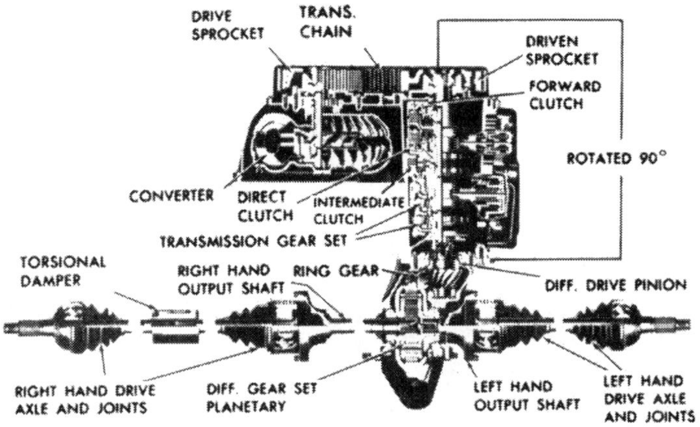

Recovery was rapid, but that was small compensation for any owner who wanted to come down the Alps at speed.

Despite the GM claims of springs that were far stiffer than the normal American rates, we decided that although roll was well controlled, we would like stiffer damping of the front suspension, which tended to "heave slowly up and down in a nautical fashion most of the time".

The car's turning circle was described as "rather clumsy" at more than 40 feet, but credit was given for the combination of a relatively small steering wheel and high gearing.

Autocar criticized the lack of stowage inside the car, just a small locker in the facia and no shelf behind the rear seats.

"All in all, this Oldsmobile is a sort of dream car that lives up to its exotic appearance in most respects and that in one blow destroys any illusions about front-wheel drive being unsuitable for very large and powerful vehicles," the report concluded.

In the USA, the reception was just as enthusiastic. Our Detroit correspondent Roger Huntingdon echoed many enthusiasts when he said that he had fallen in love with the car on its announcement and had ordered one right away. There was nothing else remotely like it on the American market either in styling, concept or engineering. He enthused about straight-line stability in cross winds, traction in the mud and snow of a Michigan winter and moaned about the brakes. The interior ventilation system was praised (*Autocar* seems to have taken through-flow ventilation as read) but Roger said the ride was definitely too harsh. "The less I feel of the road under me, the better I like it," he commented, probably mirroring the attitude of many Americans.

The Toronado was successful, in that it gave Oldsmobile a top of the line Grand Touring Coupé that was the talking point of the world. But front-wheel drive was never used in other cars in the Oldsmobile range. It was heavier and more expensive to manufacture than conventional layouts, although it was capable of taking even more power, as shown by the application to the Cadillac Eldorado with its 8.2-litre engine. Increasingly, emphasis on emission and safety requirements pulled engineers away from pure engineering experiments like the Toronado.

However, the car has had a long and successful career with GM and is still being produced. It has become bigger, coarser and less eye-catching over the years. But most importantly it laid the groundwork for the future of front-wheel drive in General Motors. It is still the *only* front-wheel-drive car, with the Cadillac, in the whole of the General Motors range, worldwide. And since GM have indicated that the future for their smaller cars will lie in front-wheel-drive, it is logical to assume that the experience with the Toronado is currently the subject of serious styling department attention. □

Oldsmobile Toronado 1977

A low-volume special has been added to the Toronado line that will surely distinguish itself as a collector's item. It's expensive at $11,132.23, but the XSR has some unusual features layered onto a super-luxury base. The back window is the eyecatcher. It's one of those marvels of chemistry that wraps around like early futuristic Studebakers, but this time *without* the pillars. It's done with a special heated rod that allows the glass to be folded with a sharp crease. In addition, the roof is notched with two hatch panels that slide in over one another to a central storage area. Fresh air is available at the flick of a switch, but the sacrifice in headroom is unfortunately substantial.

Both the standard Brougham and the XSR models are lightly restyled for 1977. Hoods and grilles are new with wide rectangles and the dominant theme. The rear is also freshly molded to produce Cadillac-like vertical bumper members.

All Toronados are now powered by a V-8 of 403 cubic inches. An onboard digital computer controls ignition timing based on signals from engine vacuum, coolant temperature and engine speed.

As the biggest Oldsmobile money can buy, the Toronado will live out its days in distinction. Make no mistake that this is a classic example of automotive abundance. The wheels are stretched to a long 122-inch separation for the ultimate in ride quality. The door handles are nearly 80-inches apart to give six lounge-level elbow room. The car is a solid 2.5-tons of wraparound comfort. Historians will certainly look back and view this front-drive chariot as one of the most interesting deviations of traditional American luxury.

Manufacturer: Oldsmobile Division
General Motors Corporation
Lansing, Michigan 48921

Base price: $8134-11,132
Vehicle type: front-engine, front-wheel-drive
Body styles available: 2-door coupe

DIMENSIONS
Wheelbase ..... 122.0 in.
Track, F/R ..... 63.6/63.5 in.
Length ..... 227.5 in.
Width ..... 80.0 in.
Height ..... 53.2 in.
Curb weight ..... 4750 lbs.

SUSPENSION
F: ..... ind., unequal-length control arms, coil springs, anti-sway bar
R: ..... rigid axle, four trailing links, coil springs

BRAKES
F: ..... vented disc, power assisted
R: ..... drum, power assisted

ESTIMATED EPA FUEL ECONOMY
City ..... 13 mpg
Highway ..... 19 mpg

ENGINES					TRANSMISSIONS			
Type	Displacement, cu in	Fuel system	Horsepower	Torque, ft-lbs	3-sp man	4-sp man	5-sp man	3-sp auto
V-8	403	1x4-bbl	200 @ 3600 rpm	330 @ 2400 rpm				X

Oldsmobile Toronado 1978

By now you should know that a Toronado is something more than a Kansas whirlwind spelled sideways. It is the largest Oldsmobile that your hard-earned dollars can buy—and it will take a lot of them to do that. You can get an idea of this car's personality by looking at the base equipment. It comes standard with front-wheel drive; 403-cubic inch V-8 engine; automatic transmission; steel-belted radial tires; power windows, brakes and steering; and air conditioning. For 1978, you can order three new leather/vinyl trim options.

The Toronado was the first car to use onboard computers to control engine performance. It's equipped with a microprocessor that Olds calls MISAR which monitors and adjusts spark timing as engine vacuum, water temperature and engine speed fluctuate. You can, of course, bask in fresh air and sunlight at the flick of a switch that opens the Toronado's unique double-panel sunroof of sheetmetal and tinted glass.

Manufacturer: Oldsmobile Division
General Motors Corporation
Lansing, Michigan 48921
Base price: $8899
Vehicle type: front-engine, front-wheel-drive
Body styles available: 2-door coupe

DIMENSIONS
Wheelbase ..... 122.0 in
Track, F/R ..... 63.7/63.6 in
Length ..... 227.5 in
Width ..... 80.0 in
Height ..... 53.2 in
Curb weight ..... 4750 lbs

SUSPENSION
F: ..... ind, unequal-length control arms, coil springs, anti-sway bar
R: ..... rigid axle, 4 trailing links, coil springs

BRAKES
F: ..... vented disc, power-assisted
R: ..... drum, power-assisted

ESTIMATED EPA FUEL ECONOMY
City ..... 13 mpg
Highway ..... 19 mpg

ENGINES					TRANSMISSIONS			
Type	Displacement, cu in	Fuel system	Horsepower	Torque, ft-lbs	3-sp man	4-sp man	5-sp man	3-sp auto
V-8	403	1x4-bbl	190 @ 3600 rpm	325 @ 2000 rpm				X

Printed in Dunstable, United Kingdom

91224062R00060